Best wishes to my friend a
Lynch associate Jack Dierks.

Enjoy,

Joe Hill

Oct. 1983

SOME EARLY BIRDS

Joe Hill
Captain, USN (Ret.)

THE NAVY FLYER'S CREED
I AM A UNITED STATES NAVY FLYER.
MY COUNTRYMEN BUILT THE BEST AIRPLANE IN THE WORLD AND ENTRUSTED IT TO ME. THEY TRAINED ME TO FLY IT. I WILL USE IT TO THE ABSOLUTE LIMIT OF MY POWER.
WITH MY FELLOW PILOTS, AIR CREWS, AND DECK CREWS, MY PLANE AND I WILL DO ANYTHING NECESSARY TO CARRY OUT OUR TREMENDOUS RESPONSIBILITIES. I WILL ALWAYS REMEMBER WE ARE A PART OF AN UNBEATABLE COMBAT TEAM — THE UNITED STATES NAVY. WHEN THE GOING IS FAST AND ROUGH, I WILL NOT FALTER. I WILL BE UNCOMPROMISING IN EVERY BLOW I WILL STRIKE. I WILL BE HUMBLE IN VICTORY.
I AM A UNITED STATES NAVY FLYER. I HAVE DEDICATED MYSELF TO MY COUNTRY, WITH ITS MANY MILLIONS OF ALL RACES, COLORS, AND CREEDS. THEY AND THEIR WAYS OF LIFE ARE WORTHY OF MY GREATEST PROTECTIVE EFFORT.
I ASK THE HELP OF GOD IN MAKING THAT EFFORT GREAT ENOUGH.
JOE HILL

SOME EARLY BIRDS: the memoirs of a Naval Aviation Cadet, 1935-1945

Joe Hill
Captain, USN (Ret.)

ISBN 0-89745-043-4 hardback
ISBN 0-89745-044-2 paperback

Sunflower University Press®
1531 YUMA (BOX 1009) • MANHATTAN, KANSAS 66502, USA

Cover Design By

Robin Higham with Lori Daniel

Book Design and Layout By

Lori Daniel with Robin Higham

Table Of Contents

Acknowledgements

I wish to thank every pilot who has had a hand in making this book possible. I have tried to tell accurately what has been reported to me, but I have had to rely on my memory to recall first-hand accounts of what has happened to some of these pilots. I will assume no liability for errors of omission or commission. Although the stories here represent only about ten percent of more than 1,200 pilots in our group, they do indicate what has happened in the lives of many of them, and bring back fond memories about the early days of flight training. During the last three years I have spent more hours than I have as a pilot (6,000) in assembling stories for the book, and putting them in some kind of readable fashion, but it has been a rewarding experience which I hope will bring some moments of pleasure to all who read it. I plan to donate ten percent of any royalties I receive to the Naval Aviation Museum Foundation in Pensacola, Florida, and ten percent to *Aerospace Historian* in Manhattan, Kansas, which is making publication of SOME EARLY BIRDS possible.

God bless all Naval and Marine cadet trained pilots!

Foreword

These are stories about Navy and Marine pilots who were trained as Aviation Cadets before World War II. Their history began in April 1935 when Congress passed Public Bill 37, creating a new rank in the Navy and appropriating funds for their flight training. Applicants for this cadet program were educated during the Great Depression, and most of them worked and saved to help pay for their schooling. They came from every State in the Union, and most of them were college graduates. They were the cream of young American manhood and very near physically perfect. They had great visual acuity and precise depth perception. They were hardy, quick-witted, intelligent, and had the spirit to accept most any challenge. They had one common thread, "A burning and intense desire to learn how to fly."

The book is dedicated to "Those Navy and Marine Aviation Cadets who gave their lives in the service of their country, and to those who have passed on whose stories may never to told."

About The Author

Born in a small Southwestern Indiana town in 1917, he completed his elementary and high-school education there, then received a BS degree from Indiana University School of Business Administration with a marketing major in 1939. Having successfully passed the Navy flight physical examination on campus in February of that year, he was inducted into the U.S. Naval Reserve as a Seaman 2nd class in June at the Naval Reserve Air Base, St. Louis, Missouri. Soloed after ten hours flight instruction, he was appointed an Aviation Cadet with orders to Pensacola, Florida, in September 1939. There he received the coveted "Navy Wings of Gold," and was commissioned Ensign, USNR, in July 1940, with orders to report to Patrol Squadron 23 in Pearl Harbor, Hawaii. He accepted a regular Navy commission as Lieutenant, USN, in January 1943 and was placed on the naval officer retired list as a Captain after more than twenty years service in November 1959. He was then employed with Merrill, Lynch, Pierce, Fenner & Smith as an Account Executive and Assistant Vice-President until retirement from that firm in January 1975.

List Of Illustrations

PART I
LEARNING HOW TO FLY

Chapter One
Time Of Decision

Naval aviation was born in November 1910 when Glenn Hammond Curtiss, a leading designer and manufacturer of aircraft, loaned one of his planes and his star pilot, Eugene Burton Ely, to the Navy to demonstrate the feasibility of launching aircraft from a ship. A launching ramp was rigged on the bow of the cruiser *Birmingham* 25' wide and 85' long, and sloping downward at a considerable angle to give the plane a faster start. The cruiser was anchored in Hampton Roads, Virginia, on 14 November 1910. The weather was far from ideal, as there was light rain, some fog, and very little wind. Ely used a Curtiss "pusher" biplane with an eight cylinder engine of 60 horsepower, and a top speed of less than 60 miles per hour. At full throttle he went down the ramp, off the bow, and all three bicycle wheels and the canvas lower wing hit the waves before he became airborne and gained flight control. But he flew the two miles to Norfolk at an altitude of 150' and landed safely at Willoughby Spit.

In January 1911 the Navy scheduled a second aircraft demonstration aboard ship, this time using the battleship *Pennsylvania,* anchored in San Francisco Bay. Again Ely was the pilot, and he planned to fly to the ship, land aboard, then take off and fly back to the field where the flight had originated. The Navy approved the experiment, but would not pay for the cost of building the necessary platform. Although Curtiss wanted him to land and take off on water next to the ship, Ely was determined to carry out his own program. He planned, designed, built, and paid for the platform over the bow with $1,200 of his own funds. To arrest the plane, he used a series of ropes attached to sand bags so that his hooks would catch them up while landing. Navy engineers did erect canvas nets along the platform to catch the plane if it swerved. Ely's safety gear consisted of two bicycle inner tubes looped around his shoulders to give bouyancy in case he ditched, and a leather football helmet. On 18 January 1911 he took off and flew to the anchored *Pennsylvania,* which was "weather-cocked" into the wind. He flew almost even with the bridge before he cut his motor and dropped to the platform where his hooks caught up the ropes and stopped the plane. Later that day, he used the entire platform for take off and flew back to the field ashore.

One month after Ely's round trip, Curtiss demonstrated the feasibility of landing and taking off in the ocean near a ship with an aircraft on floats. As a result of these demonstrations, the Navy ordered its first aircraft.

Although Curtiss and Ely were civilian pilots, they worked closely with the Navy, and have often been referred to as "The Father and Son of Naval Aviation." Both were among the first twelve selected for enshrinement in the Naval Aviation Hall of Honor in the museum at Pensacola. Curtiss built the first Navy aircraft, the A-1 Triad, and trained the first naval aviator, Lt. Theodore Gordon Ellyson, a 1905 Naval Academy graduate. All naval aviators since Ellyson have been designated with consecutive numbers.

Most of the early aviators in the Navy were trained by the civilian test pilots who worked for the aircraft manufacturers. But once the aviators were thoroughly checked out in their plane types, they learned to fly by teaching each other. On one of these training flights out of Norfolk, Virginia, Lt. John Towers was teaching Ensign Billingsly how to fly, with Billingsly at the controls. During one of their manuevers, at less than 500', Billingsly was thrown from the aircraft and was killed. Towers could not gain control of the plane before it crashed into the Chesapeake, but he clung to a wing strut after the crash and was rescued by a nearby boat crew who had witnessed the entire episode.

As a result of this fatal accident, Lieutenant Towers designed and developed the first Navy parachute, and thereafter parachutes were required equipment on all flights. Towers was a 1906 Naval Academy graduate who was designated naval aviator # 3. He was a foremost pioneer in naval aviation at all levels during his career, was Chief of the Bureau of Aeronautics 1939-42 during the tremendous build-up of naval aviation before

World War II, and is one of twelve enshrined in the Hall of Honor.

Because of their invaluable contributions to naval aviation in the early days of flight, I believe the other Hall of Honor enshrinees should be recognized here. In addition to Curtiss, Ely, and Towers, there was T. G. Ellyson, aviator # 1, who was personally trained by Curtiss, free of charge. P. N. L. Bellinger, # 4, was a pioneer in all phases of aviation and went on to become one of the naval leaders in World War II. Aviator # 5 was the first Marine pilot, A. A. Cunningham, and is recognized as "The Father of Marine Corps Aviation." He also was the first man to be catapulted from a warship underway. Godfrey Chevalier was # 7. He commanded the Northern Bombing Group in France in World War I and was commander of the original aviation detachment assigned to the first carrier, the *Langley*. Aviator # 13 was H. C. Richardson, an aircraft designer who supervised the construction of the famous NC series aircraft. He was also a pilot in the NC-3 which attempted an Atlantic crossing in 1919. A. C. Read was # 24 and was command pilot of the NC-4, the only one of the NCs to complete successfully the 1919 Atlantic crossing. W. A. Moffett was the first Naval Aviation Observer and the first Chief of the Bureau of Aeronautics (1921-33). He was largely responsible for keeping aviation in the Navy during the lean years in between World Wars I and II. Floyd Bennett was the first of a long list of Aviation Pilots (outstanding enlisted men selected for flight training) in the Navy and was co-pilot with Richard E. Byrd on the first flight over the North Pole in May 1926. The last of the twelve men to be enshrined in the Naval Aviation Hall of Honor thus far is Richard E. Byrd. He graduated from the Naval Academy in 1912 and became naval aviator # 608. He was the first pilot to fly over both poles, and devoted more than thirty years to exploration of polar regions.

Six individuals each year will be selected for enshrinment in the Naval Aviation Hall of Honor. Nominees may be uniformed or civilian personnel, but they cannot be nominated as long as they are employed by the government. The Deputy Chief of Naval Operations (Air Warfare) will appoint a selection committee each year to determine which six nominees will be selected. Enshrinees normally will be selected from individuals recognized for sustained superior performance in or for naval aviation, from those who have made superior contributions in its technical or tactical development, and from those recognized for unique and superior flight achievement in combat or noncombat operations.

The early aviators in the Regular Navy and Marine Corps were well aware that the Regular Establishment could not provide enough aviators for the fleet. Thus, in the early 1930s they founded the Reserve Student Officer program under which aviators were trained as enlisted personnel, then commissioned Ensigns, USNR, and sent to the fleet for a year.

Most of those recruited into the new RSO program were college graduates who wanted to learn to fly, and most of the recruiting was done from the twelve Naval Air Reserve Bases. The bases covered the country geographically so that in the West they were at Seattle, Oakland, and Long Beach; in the Midwest at Kansas City, St. Louis, Minneapolis, Chicago, and Detroit; and in the East at Boston, New York, Philadelphia, Washington, and Miami. These bases were set up to screen and physically examine applicants, and between 1932 and 1935 several thousand were processed. However, the program was suspended in the Spring of 1933 when President Roosevelt declared that he would balance the budget. Although thousands were processed, only about 100 finished their flight training as enlisted men and were sent to duty as Reserve Ensigns. Politically, the program was a great success in that it provided a few jobs during the Great Depression, and it covered all sections of the country. However, the number of pilots trained was far from adequate. Yet the program did establish a nucleus of fine young Reserve aviators for the Navy. But it was not completely successful, because one year of active duty was not enough to make these Reserves an effective force with the fleet.

With this in mind the aviators in the Regular Navy did a tremendous job of lobbying to get a new Reserve pilot training program authorized, and on 15 April 1935 Congress passed Public Bill 37. This authorized appropriations for the Naval Aviation Cadet program, and established a new officer rank in the Navy. The lobbying feat seems incredible, because just two years earlier the Navy could not afford to send Academy graduates in the lower half of the class to active duty. However, the program was very popular with Senators, Congressmen, and the Roosevelt administration, because it provided a large number of attractive jobs during the depressioned economic times.

What kind of men did the Navy wish to attract in this new Reserve Officer rank? They wanted the very best, "the cream of the crop," and they considered only college graduates or those in their senior year. They had to be at least five feet, six inches tall, and they could not be over six feet, four inches. They could not be under 20 years of age, nor older than 26. They had to have most of their teeth except wisdom teeth, and they had to be in good condition. They could not be overweight or underweight for height, and they had to present a good over-all appearance. They had to be clean shaven and could not have beards or mustaches. They had to have 20/20 vision and precise depth perception. Their hearing had to be acute, and they had to be mentally alert. In fact, they had to be almost physically perfect and agree to remain unmarried for four years. Most applicants were screened and examined at one of the twelve Reserve bases, or by recruiting teams from those bases that went directly to college campuses to give examinations.

What inducement did the Navy offer? Just ten hours of dual flight instruction at one of the Reserve bases, as

Seamen 2nd Class, USNR, the lowest rating in the service! The Reserve bases were referred to as Elimination bases, or E bases, and aptly so, because about 95 percent of original applicants were eliminated before they ever became cadets. The pay for a Seaman 2nd Class was less than a dollar per day, but there was a food allowance of thirty dollars a month if there was no established mess on the base. Seamen learned quite a bit about flying and the Navy at E Bases, as about fifteen hours of training on the ground was required for each hour of flight training. They learned a new glossary of Navy terms that had been around the British and American navies for hundreds of years. They learned about parachutes, safety equipment, power plants, flight instruments, and navigation, but mostly they learned about washing down airplanes and cleaning heads (toilets).

Once Seamen 2nd Class finished thirty days of E Base training and had safely soloed, they were commissioned Aviation Cadets, USNR, and assigned to a flight training class at Pensacola for one year. The pay at Pensacola was $75 per month plus room and board. When Cadets had successfully completed all phases of training they were designated Naval Aviators with consecutive numbers by date and then by class standing and were sent to the fleet for two years. Their pay was increased to $125 per month and they were given a $30 per month food allowance. They were furnished living quarters aboard ship, or lived in bachelor officers quarters while ashore. They were assessed a monthly mess bill aboard ship, or when living ashore, but the $30 allowance was more than enough to cover mess bills. After two years in the fleet, they completed the final year of their contract at one of the twelve Naval Air Reserve Bases. They agreed to remain unmarried for the entire four years of the contract, and they were given a $500 per year bonus for each of the three years as a Naval Aviator. And this was all good money in those days when a college professor made perhaps $3,000 per annum and a school teacher $850 for nine months!

To get a sufficient number of applicants for the cadet program, recruiters had to do an outstanding job, because only one in twenty was accepted for E Base flight training. Of those accepted, an average of 30 percent were "washed out" and failed to solo. Of those commissioned and sent to Pensacola, 40 percent failed to complete the course and did not become aviators.

The first aviation cadets reported to Pensacola in July 1935 and were assigned to class 81 C. In the first year group a new class reported every month for seven months in classes numbered 82 through 88 C. There were 69 in the first class and 65 to 70 in each of the other classes. Although more than 500 cadets started in training that year, fewer than 300 were designated aviators and sent to the fleet. In the next three years more than 1,200 started training, but fewer than 700 went to the fleet. In all, about 1,000 Reserve Cadets had gone to the fleet by the Summer of 1939.

Beginning with our flight class (130 C) in September 1939, more than 100 cadets were assigned to each class. As the country got closer to war, flight training activity was stepped up considerably, so that more than 1,500 cadets started training at Pensacola between the Summer of 1939 and the Summer of 1940. More than 1,000 cadet aviators got to the fleet that year. As the number of cadets increased the following year, Pensacola could not accommodate all of them, and thus new training bases had to be established at Jacksonville, Florida, and Corpus Christi, Texas. By the time the Japanese attacked Pearl Harbor on 7 December 1941, the three bases were turning out thousands of reserve aviators each year.

The scope of this book is limited to those cadet "Early Birds" who received their wings and went to the Fleet before 7 December 1941. The book was conceived during a reunion of several hundred of these gentlemen at Pensacola in November 1979. My good friend and Pensacola classmate, Frank "Delo" DeLorenzo, was chairman of the committee that sponsored the reunion, and he and his wife, Helen, had compiled a list of more than 1,000 ex-Navy and Marine cadets who are living today, complete with addresses and telephone numbers. I wrote a letter to every one on the list, announcing my intention of writing a book, and welcomed comments about their experiences. I was pleasantly surprised by their generous response, and have used many of their stories. This group of aviators proved to be the very backbone of naval aviation during the war years and was a factor in every naval operation.

One of the first stories that comes to mind is my own. My grandfather, C. M. Hill, was a "Drummer Boy" for the Union Army in the Civil War. He came to Bruceville, Indiana, in 1873 and opened a General Store. The same store is being operated today (1983) by his grandson, G. M. "Mickey" Hill. My mother and father were born in Bruceville in 1895, and I arrived in December 1917. I was six years old when I saw my first airplane. A local World War I pilot and hometown hero by the name of Joe Utt worked and saved until he had enough money to buy an army-surplus JN-4 "Jenny." It was a biplane complete with wing skids, tail skag, and no brakes. He began barnstorming around the Midwest in 1924, and every time he and the airplane needed a rest, they came back to Bruceville to land in the corn field south of town. Joe had an agreement with Jim Ashby, who owned the corn field; he allowed Joe to put a windsock at one end of the field so that he could come and go as he pleased.

Joe was an outstanding aircraft mechanic, and he performed all of his own maintenance work. If the plane needed parts, he would have the parts when he returned from his barnstorming. He would work on the plane a few days, take some friends for a ride while he was testing it, and be off on another tour. The Jenny had two open cockpits, and the Indiana winters were so harsh

The Wayne McIntyre Orchestra. Hill seated next to lady vocalist playing for Presidents' Ball, Brazil, IN, 30 January 1936.

that it was too cold for barnstorming once Fall arrived, so Joe would bring his plane back to Jim Ashby's barn where he stored it until he brought it out the following Spring.

I was nine years old in May 1927 when Lindy flew across the Atlantic to Paris, and I remember we followed his progress by radio all of the way. What a hero! About this time I began to mow some lawns and stack some wood, in order to save money and get a bicycle. This is a poem Grandfather wrote about it:

If you have any wood to stack
Or any lawns to mow
I know a little cracker jack
Little curley headed Joe

He'll do your work and do it right
And he is not so very slow
For he works with all his might
This curley headed Joe

You'll not find a better hand
No matter where you go
Than this boy. He's got the sand
This curley headed Joe

When I saved enough, I got my bicycle — and a paper route. When I outgrew the bike and the paper route, I worked in a filling station in the daytime and played saxophone in a dance band at night. With the money I saved and some help from Dad, I managed to enroll at Indiana State University in the Fall of 1935. I played sax with Wayne McIntyre's band all that year and the following Summer. The next Fall I enrolled at Indiana University as a pre-law student and joined Delta Upsilon fraternity. One of the most lingering impressions about fraternity life was the Monday night "board" meetings. Although I was a college sophomore, I was a freshman in the fraternity, and freshmen were not allowed to play poker or gamble with upper classmen. I had been playing poker since I started mowing lawns, and the rules did not stop the upper classmen from inviting me to play when they needed another hand. We were all good friends and brothers until Monday night, when I got the full benefit of their paddles at the "board" meeting. It was especially painful if I had been a winner during the week. We also had a few bridge games for a tenth of a cent a point, and they put the hurt on me for those.

The Great Depression was in full bloom, and I knew Dad did not have much money. However, when he took me to Indiana University for "rush" week, he handed me a check book and said, "Son, if you ever really need something, write a check for it and I will cover it." I truly appreciated this confidence in me, and I doubt if I wrote more than $1,000's worth of checks total in all three years. I always had my saxophone with me, and I managed to pick up a few small jobs on weekends, but I was not a member of any regular band. I worked for meals in the Tri Delt Sorority during all three years, and I was "Head Waiter" for the last two years. With this job went the responsibility for a coal-fired hot water heater to assure that the girls had hot water all year; I was paid $5.00 a month for this duty. During the cold winter nights I would have to go to the sorority and re-stoke the fire about midnight, because the coal would not last until breakfast time.

Although I was a pre-law student my first two years at Indiana, I abandoned the idea of Law School in my senior year and decided to get a degree in Business Administration as a marketing major, because with Hitler's rise to power in Europe it looked like we were surely headed for war. This required a six-hour overload of subjects during my final year, but I made better grades than I had made in any of the first three years.

I had plenty of girl friends in High School and college. I was rather serious about one in High School days, but realized that there wasn't any way that I could make enough money to take care of two people during the Depression. A serious relationship that could lead to marriage was out of the question. This was brought out loud and clear during my final college semester when the best job offer was $100 a month with an insurance company.

One cold and blustery afternoon in late February 1939, fraternity brother Tom Groscop and I were walking through the "Commons" (a wooded area in the center of the IU campus), when we noticed several Navy recruiting placards. They invited us to "Join the Navy and learn how to fly. Follow the arrows and take a flight physical examination." This seemed like a very good idea, as we had been out the previous evening and were dubious about the current state of our health.

We began the examination in the early afternoon, and it was after 6 p.m. when we finished. One of the final steps in the exam was to have our eyes dilated, and if the pressure was OK, we were to see a 20-minute film called "Navy Wings of Gold." We passed every test with flying colors, watched the movie, and learned that we were the only two people examined that day who did so. We were impressed with the thorough and precise manner in which the test was conducted, and the examining team was impressed with our good health. They asked if we knew of a good restaurant in town. We suggested Gus Lycas, the Greek's, for he had the best spaghetti and beer in the city, and soon we were all off for another full evening. We were ready to sign any papers they sent us, and by the end of the first week in March, we had them completed, signed, and sent back to St. Louis with our character references. Under the terms, we were to report for duty as Seamen 2nd Class to the Naval Air Reserve Base in St. Louis, Missouri, on 15 June 1939.

Between the time we finished the examination and the time we finished signing the papers, Tom and I did a lot of thinking and talking. We wondered if we were doing the right thing. Neither of us had ever considered military service, or seriously considered flying. However, we both were certain that Hitler was about to start a war, and believed that the sooner we got some kind of officer training the better off we would be, even if it meant flying. Furthermore, the best part of the program was that we were committing ourselves for only thirty days. If we decided during that period that we did not like to fly, we could pack it in and go home and start looking for a job.

The one thing we failed to talk about was the reaction our parents might have when we told them about our plans. There were not many airplanes flying in our hometowns in those days, and flying was considered very dangerous. Also the reputation enjoyed by the Navy in the Midwest, left something to be desired.

We went home for Spring vacation the first week in April, broke the news, and all hell broke loose. Tom's mother almost had a heart attack, and he believed she would surely die if he carried out his contract. He promptly wrote the Navy, explaining the circumstances, and they cancelled his orders. When I told my parents, they were far from pleased with my decision, but they encouraged me to give it a try if I was sure that I wanted to learn to fly. After all, I pointed out, the trial period was only thirty days, and I could quit any time during that period if I did not like it. My decision stood, I liked it, and went on to fly almost 6,000 hours for the Navy. My father showed his mettle in allowing me to try aviation, because during his entire life he would not go near an airplane (he thought it was too dangerous). My mother, however, has flown many times.

In looking back, I guess the two most important forces motivating me toward aviation were (1) Joe Utt, the barnstorming World War I hero, who operated his plane out of the corn field South of Bruceville, and (2) Charles Lindbergh's solo flight from New York to Paris in 1927. These two events made a strong imprint on my subconscious, because I had not been in the air five minutes on my first instructional flight when I realized that flying was exactly what I should be doing. I had made the right decision.

Although my introduction to the Navy and naval aviation was quite simple, many of my cadet friends were not as fortunate. Most of them were physically examined at one of the twelve Naval Air Reserve Bases or at a Naval District Headquarters. Some of them had to take more than one physical examination before they were accepted, and the odds after that were still stiff as noted earlier. Thus it may be said that the cadets who got to the fleet were something special.

Walter A. Hibbs, who now lives in Carmel, California, was Commander of the Banner NROTC Company during his senior year at the University of Washington, in Seattle, in 1933. He had learned a lot about the Navy in the NROTC program, and he dearly loved it. He was commissioned Ensign, USNR, in June 1933. However, before he was sent to active duty, he learned about the Reserve Student Officer program from a Marine friend at Pensacola. His Marine friend had been so enthusiastic about the program that Walter decided to apply for it. He resigned his commission and was accepted into the Reserve Student Officer program. Learning how to fly sounded like fun, and it was a better paying job than he had ever had.

NAS Sand Point, Washington, gave an unofficial

Consolidated NY-2. (Naval Aviation Museum)

ground school as a means of RSO selection, and students washed down airplanes on weekends. Sometimes they were rewarded with an occasional ride in the back seat of a Consolidated NY-2 and given some flight instruction. However, before any of the RSO selectees received orders to flight training at Pensacola, President Roosevelt cancelled the program. The next year, Walter moved to Port Townsend, Washington, and had a job as a chemist with a paper company. He could not get to Sand Point for weekend training and RSO selection in 1934, but he learned about the aviation cadet program in 1935 and decided to apply for it. The cadet selection board gave preference to previously selected RSO students, so Walter was one of those chosen for the first cadet training class at Pensacola in July 1935.

Flight training as an aviation cadet was not his first choice as an occupation. His father was a doctor, and his girl friend's father was a doctor, and his first choice was medicine. But he did not have funds for medical school, so he accepted cadet flight training as a fun second choice, and he has never regretted it. His motivating force, the Marine Reserve Student Officer, served his year in the fleet and went to work for Boeing as a test pilot. Unfortunately, he was killed a year or so later while testing the Boeing "Stratoliner."

Another one of the first-year cadets (1935-36) with an interesting motivation story is Howard Kinchloe, who now lives in Tucson, Arizona. He was born in Wichita, Kansas, which while Howard was growing up in the late 1920s and early 1930s was referred to as "The Air Capital of the World." Cessna, Stearman, Swallow, and Travelair were all major aircraft companies. There were seven air fields in the area used for testing planes.

During this period Howard was in high school, and he spent all of his spare time around the aircraft plants or the nearby airfields. He had his first ride in 1929 in a Travelair owned by his cousin. It was piloted by "Swede" Christopher, who was the top salesman for Travelair. Howard worked for Swede after school and on weekends and during the Summers of 1930 and 1931. Swede rewarded him by giving him flight lessons, and he received his student's flying permit in 1932. Also that year, Howard worked on Jimmy Doolittle's "Super Solution," and Swede took him to the Cleveland Air Races where he watched Jimmy perform. He also met such aviation pioneers as Roscoe Turner, Jimmy Mazlip, Jimmy Wadell, Keith Reider, and Beey Howard, and he saw the Granville Brothers' "Gee Bee" win the Thompson Trophy.

By the time Howard learned about the aviation cadet program, he had quite a few hours as a pilot, and he knew a lot about airplanes as he had helped a friend rebuild a Stearman that had cracked-up. It took him a week to complete the physical examination at 3rd Naval District Headquarters. During that week he learned that the minimum weight for his height was 135 lbs., and he weighed only 128. However, he ate enough bananas and drank enough water to make the 135 lbs., and he successfully passed the exam. He received his E Base training at Floyd Bennett Naval Air Base, N.Y., soloed there, and went to Pensacola in November 1935 for flight training in class 85 C. Unlike Walter Hibbs, flying was Howard's first choice, and the only occupation he ever considered. After his service as a cadet, he became a chief test pilot for Vultee in their Louisville plant during World War II.

In the second-year group of cadets (1936-37), Walter C. Michaels, who now lives in Ojai, California, became interested in flying when he was 11 years old. He saw every aviation movie that came to town and read everything about aviation that he could find. He lived in New York City all of his life, and he was most impressed when Lindbergh took off from Roosevelt Field and made it all the way to Paris. He graduated from New York University with a B.S. in Mechanical Engineering (Aeronautical option) in 1936. He learned about the aviation cadet program from a notice on an NYU bulletin board, and because he had not had a decent job offer, he decided to apply for it. He was told to report to the 3rd Naval District Headquarters for a flight physical examination. He reported for the exam with 22 other applicants, but Mike was the only one who passed. Late in the Summer of 1936, he was inducted into the Naval Reserve as a Seaman 2nd Class and started his flight training at Floyd Bennett Field. There were 24 Seamen in his class and 20 soloed and went to Pensacola. Mike was one of the lucky 10 that got their wings, and he was designated Naval Aviator # 5,500.

David M. Burns, of Whispering Pines, N.C., says he had never even thought about an airplane until he got his first plane ride in the early thirties. The price for each passenger was one cent per pound for the ten-minute ride. David (nicknamed "Doc" he says because his initials were M.D., but not in that order) weighed only 135 lbs., and the other passenger was not much heavier, so the price was right. He graduated from the University of Texas in the Spring of 1936, and he had not been impressed with any job offers up to that time. He had not thought much about aviation since his first plane ride, until he learned about the Navy cadet program. He found that he could take the flight physical examination in Dallas, and so he reported in on the appointed date with about fifty others and was one of three that passed it. He was inducted in September 1936 at the Naval Air Reserve Base in Opa Locka, Florida, and began his flight training there. His class of twelve lived in tents (two to a tent) and slept on army cots with issue bedding (no mosquito nets). They ate their meals in the general mess. They received ten hours dual flight instruction in Curtiss Fledglings with tail skags and no brakes. All landings and takeoffs were made from grass fields. "Doc" was one of the lucky three that soloed and made it to Pensacola in class 97 C.

From the next year group (1937-38) several ex-cadets gave some interesting stories about their motivation into the Navy and aviation. Ed Kiem, of Pensacola, says he used to hang around the Newark Airport, N.J., when he was in high school. He would go there after school and watch the celebrities come off the transports. He marveled at the mobility the planes afforded. He had read in the morning paper the speech Will Rogers gave in Oklahoma City the night before, and here the famous man was getting off a TWA the next afternoon in Newark.

Ed says his first motivation toward the Navy came in the Summer of 1934 in New York City. The battleship *Mississippi* was anchored in the stream of the Hudson River and was holding "Open House" during the Fourth of July weekend. The Navy was offering free rides from the dock to the ship and back in fifty-foot motor launches. Ed was impressed with the immaculate dress of the boat's crew and the personnel aboard the ship, and with the "spit and polish" he observed on the tour. Little did he realize at the time that the *Mississippi* would be his home as a pilot for 2½ years at a later date.

Ed was attending Worcester Polytech on a scholarship during his first two years and was living in a flat for $15 a month and parking cars in a garage for $15 a week. He could save no money toward his junior year, and without the scholarship he could not complete college. He read about the scheduled visit of a Navy recruiting group that was looking for prospective flight training students with engineering background, so he decided to try his luck. After hearing the recruiter's pitch and seeing the movie "Navy Wings of Gold," he was sold on the program and began searching for the place to take a flight physical. He and his roommate made an appointment to take the examination at the District Headquarters in Boston. Ed passed, but his roomie "busted" his hearing test. Two weeks later Ed reported to the Naval Air Reserve Base in Squantum for an interview, was accepted into the program, and was ordered to report for induction on 15 June 1937. He was one of the few accepted into the program with only two years of college up to that time, and when he was sworn in as an Aviation Cadet he was the youngest ever, just three days past the minimum age of twenty. Ed reports that the weather was so bad during the 30-day training period, that all eight seamen were given their solo flight checks on the last day. They all soloed and went to Pensacola, but only two received their wings and were designated Naval Aviators.

Charles Harbert, another cadet in the 1937-38 year group, got in the Navy because he was starving to death teaching school in West Virginia. He began teaching in 1930 at $150 per month. His salary was reduced each year until 1933, when it was cut to $97.50 per month. It remained at that level until 1935, but then when he cashed his first monthly check it was returned with the notation, "Insufficient Funds." Charlie decided that he had better get the hell out of the teaching business, because he would not work for nothing. He had never thought about flying, nor of entering any type of military service, but when his girl friend showed him a piece in the paper about the aviation cadet program the Navy was offering, he decided to look into it. He found that he could take the flight physical examination at the Naval Air Reserve Base at Anacostia. He reported one Saturday morning with about sixty other applicants, and he was one of the lucky four that passed. There were nine in training altogether, and eight in his class soloed and went to Pensacola. He was the only cadet that received his wings and went to the fleet, and he considered himself to be most fortunate when he looked back on his night "bail-out" from a fighter plane.

Charles Ingalls was another in the 1937-38 year group who caught the flying bug early in life. In 1930, while working for the highway department in the Summer, he saved enough money to take flying lessons. He loved to get into the air, and before the Summer ended, he had passed his solo flight check. After two years of college he applied for Army flight training, but was turned down because of a severe case of acne. He went back to Auburn for his final two years, but during the last semester he learned about the Navy's cadet program. He took the flight physical and was one of the few that passed it. Two weeks later he received a letter stating that he had been selected into the program as 16th alternate, and before he could get his aeronautical engineering degree, he received a notice to report to Opa Locka, Florida, for E Base flight training. He was ecstatic, for he was going to do exactly what he wanted to do, and would be paid for it. The pay was not all that bad when one considered that the top engineer in his class was offered only $125 a month, and the top one-third of the class was offered about $90 a month. The bottom two-thirds of the class did not get any job offers, so they had to go home and shoot pool. When Charles reported to Opa Locka, his training class was one-half finished, but he had no problem completing the work or passing his solo check, because he had done it several times as a civilian.

Two other cadets in this year group had considerable experience in and around airplanes before they ever considered the Navy's cadet program. One, Howard Dostal, now of Redondo Beach, California, had been excited about airplanes since he was a little boy. One day while riding his tricycle in front of his home, he saw his first plane flying very low, and it scared him half to death. At 16, he joined a Boy Scout Glider troop and helped build a glider. The troop was sponsored by Alexander Industries of Colorado Springs, Colorado, and its glider was called an Alexander. Howard learned to fly in the troop's primary glider, and he had his first bad moment when a fellow scout flew it right into the ground and was killed. The troop leader wisely advised all in the troop to get back in the air as soon as possible, and this they did when another primary glider became available. Howard had his first powered airplane ride, during his early college years, in an Alexander "Fly-about." He had ten hours dual flight instruction before going to Purdue University for his last two years of college. There he joined the Purdue glider club and logged ten hours in Baker Cadet secondary gliders during his junior year. During his senior year he joined the Purdue Flying Club and soloed after 2½ hours of dual instruction. He went on to get an amateur pilot's license after 25 solo hours, and a private pilot's license after 40 solo hours. His first passenger was his college sweetheart, Jeanne Beemer, whom he later married. He passed Army and Navy flight physicals, but chose the Navy because he had always been a "landlubber."

Another in the 1937-38 year group who had a lot of flight time before he became a cadet is Marine Hamilton Lawrence, now living in Coronado, California. He says Lindbergh's flight to Paris got him started, and he became president of the Galileo (San Francisco) model airplane club. He soloed in a "Moth" in 1932 while he was still in high school. When he was 21, he came into some money from his grandparents, and he blew it all on a Waco YKS five-place biplane. It took him less than a year to learn that he really could not afford to own an airplane. He learned about the Navy's cadet program at the Oakland Reserve Base and passed his flight physical there, but the Navy quota was filled. However, a Marine Reserve Captain by the name of R. C. Mangrum signed him for a spot in the Marine quota, and he went to the next E Base training class at Oakland. There were nine in his class, and six soloed and went to Pensacola. Because he had over 300 pilot hours before he ever signed as a Marine student, he had no trouble passing every flight check. He was the first student in his class at Pensacola to get his wings, and a commission. He reported to VMB-1 in Quantico, Virginia, in late 1938.

One of the most unusual motivational stories in the 1937-38 year group is that of Lyman McAboy, now of McLean, Virginia. Mac got into the aviation cadet program strictly by accident, and mostly because he was an excellent baseball player. He had finished college, and could not find a job, but he rather expected a Scout for one of the major league baseball teams to pick him out — he was that good. One of the best baseball amateur teams in the Washington, D.C., area was at the Naval Air Reserve Base at Anacostia, but had a big hole at shortstop. Mac was good enough to play any position but pitcher or catcher, so he began working out with the team as shortstop. He filled in so well that the other team members talked him into taking a physical examination for the cadet program. He took it one Saturday morning with about fifty other applicants, and he was one of the two that passed. He then became a bonafide team member and played with the team until he was sworn in as an aviation cadet and sent to Pensacola in class 107 C. He didn't get to the majors, but he made the Navy his career and amassed thousands of hours as a pilot before he retired.

In the 1938-39 year group there are several interesting stories about people who got into the aviation cadet program. One is that of John Kinney, a retired Marine officer, who lives in Portola, California. He became interested in flying when he was 12 years old. Two teen-age neighbors bought an Army surplus "Jenny," and with the help of a mechanic, they had it in flying condition in a few weeks. John watched the whole restoration procedure, and was fascinated by their tales of joyriding in the area until they "cracked-up" about a year later. John was still quite young when his uncle, a banker, bought a Jenny of his own, and gave John his first ride. That ride made him determined to become a pilot. Another uncle had been a World War I pilot who was killed just before the Armistice was signed and had named John as one of four beneficiaries of his $10,000 insurance policy. This made it possible for John to attend Washington State University, where he majored in electrical engineering. After two years of college, he applied for training as an Army aviation cadet, but was turned down when the doctor erroneously diagnosed him as being color blind. He went back to school and earned his engineering degree, then went to work for Pan American Airways as a mechanic for 26¢ an hour. After 1½ years he was making 36¢ an hour and bringing home more than $20 a week, but he was dissatisfied and wanted to learn to fly. Some of his fellow mechanics were making much more, and had saved enough to buy their own airplane. He talked them into giving him flying lessons for $7 per hour. When he was ready to solo, he took the flight physical exam for a pilot's license and passed. This was concrete evidence that he was not color blind, so he applied for Army Air Corps training again, but doctors dug out his old file and rejected him again. He then turned to the Navy, but there were no openings in his district. Navy recruiters did suggest that he try the Marine Corps. They were impressed with his credentials (electrical engineering degree and seasoned mechanic with flight experience). He successfully passed the Navy flight physical, received a leave of absence from Pan Am, and went to Pensacola in class 117 C. He and Bill Kellum were the only two cadets who received regular Marine Corps commissions in that class.

Carl Rinehart (1938-39 group) says his older brother, Clark, proved to be his motivating force. They had both finished college and discussed taking the flight physical at the same time in 1937, but Carl had a love interest at the time and decided to continue teaching school in Garden Grove, Iowa. Clark successfully passed the flight physical and was inducted as Seaman 2nd Class at the Naval Air Reserve Base at Fairfax Field, Kansas City, Missouri. He received thirty days training, soloed there, and was a cadet assigned to Pensacola class 104 C. When he came home the following Summer with his new uniform, new Plymouth, and orders to a fighter squadron on the carrier *Lexington,* Carl was envious. But he really became motivated when Clark said, "Carl, it's a good thing you didn't go to Pensacola with me, because you would not have made it through the course. You wouldn't have put up with all of the insults and rough treatment we received from instructors and check pilots. You would have punched one of them in the nose, and been washed-out completely." Immediately, Carl began checking on where and when he could take a flight physical. He passed the exam at Fairfax Airport in Kansas City and began his flight training there in August 1938. He soloed 10 September 1938, but didn't get to Pensacola until the end of the year in class 121 C.

Carlton H. Clark (1938-39 year group) was going to Western Kentucky State Teachers College (now Western Kentucky University) in Bowling Green when he first learned about the Navy's cadet program. He found that he could take his flight physical examination in Louisville, and was one of the few that passed the day he took it. He was interviewed there, was accepted into the program, and was told he would be notified when to report to the Naval Air Reserve Base in Robertson, Missouri. He reported in August 1938 with seven others in his training class. They slept on Army cots with Navy issue bedding (no mosquito nets) on the second floor of the American Airlines hangar, above the office spaces. Each Seaman had a small metal locker beside his cot; otherwise the second floor was bare. The head and shower facilities were in the Navy hangar, several hundred feet from sleeping quarters. There was no general mess. Although some may have considered the living conditions to be a real hardship, Carl said he would have gladly have paid the Navy for the flight training if he could have afforded it. He was so thrilled with learning how to fly that he was not aware of any hardships. Only two of the eight in his class soloed and went to Pensacola and they were both named Clark in 122 C, although the other one spelled his name Clarke, W. E..

Hugh S. Rogers (1938-39) graduated from the Citadel in 1938 and was an outstanding amateur boxer who fully expected one day to become a professional. However, before he could establish himself in boxing, he learned about the Navy's cadet program. He found that he could take the flight physical exam at Naval District headquarters. He successfully passed the exam, was interviewed there, and was told to report to the Naval Air Reserve Base at Opa Locka, Florida, for E Base flight training. Hugh (nicknamed Buck) says they lived in tents with wood floors and slept on cots with issue bedding and no mosquito nets. They had three meals a day in the general mess. All landings and takeoffs were made from a grass field and were controlled from the tower by blinking red or green Aldis lamps. Buck soloed without any trouble and went to Pensacola to become a Wing leader in class 120 C. He now lives in Inverness, Florida, less than 10 miles from us here in Floral City. What a small world!

Ben Troemel (1938-39), now living in St. Augustine, Florida, says, "I was determined to become a professional pilot, when at the age of 12 years, I was thrilled by Lindbergh's flight to Paris in 1927." As soon as he was old enough to begin flying lessons, he got some instruction any time he could afford it. In 1935, while a cadet at the New York State Merchant Marine Academy, he made his first solo flight after a total of 7 hours and 20 minutes of dual instruction. However, his civilian flight activities ended there due to lack of funds, and it was not until 1938 that he learned about the Navy's cadet program. He took the flight physical at the nearest Naval District headquarters, passed it, and went on to Elimination Base training where he soloed again in August 1938. He was in class 121 C at Pensacola.

Whitney Wright (1938-39), now of Boothbay Harbor, Maine, had never thought much about aviation, or the Navy, until he was a junior in Colby College in 1937. Robert V. McGee, his roommate, had become interested in the cadet program and was about ready to submit his application and quit school if accepted. However, Whit wanted to finish college and talked Bob into staying one more year so that they could submit their applications together. After graduation, they both went to Squantum, Massachusetts, for flight physicals and interviews, and were accepted into the program in the same E Base class. There were 13 in the class, and they all soloed, but McGee "busted" his physical before being sworn in as a cadet and did not arrive in Pensacola until six months after Whit was assigned to class 122 C.

Although I knew many cadets in the 1938-39 group, as they were there when I arrived, I am most familiar with my contemporaries in the 1939-40 group. Norman Brady, a Pensacola classmate who is retired there, reports that he had wanted to fly as long as he can remember, and after Lindbergh's flight he was determined to be a pilot. He made his first application for flight training with the Army in 1934, but was unsuccessful. His next try was with the Navy in 1936. He hitchhiked from Sioux City, Iowa, to the Naval Air Reserve Base in Minneapolis, Minnesota, where he was to take his physical. He passed all phases of the examination with flying colors until he weighed in two pounds under the minimum weight for his height. He was rejected on the spot, and when he asked what he could do to be sure of getting into the program, he was told to get two more years of college, put on a few pounds, and try it again. He followed that advice, and in the Spring of 1939, he and a friend, Dale Rogers, hitchhiked to Minneapolis. They reported for their physical exam at 8 a.m. one Saturday morning with 26 other hopefuls. By 4 p.m. he and Dale were the only two that passed the exam and were accepted. Two months later they were in the same E Base training class, but Dale "busted" his solo check, so Norm was the only one that made it to Pensacola in class 130 C.

Frank L. DeLorenzo, another Pensacola classmate and good friend, bought a house near the 18th green at the Pensacola Country Club in the 1940s, and he still lives there. Delo says he has been interested in flying as far back as he can remember. He built model planes as a boy and made light planes that were shot with rubber bands. He sold them for $1 each. He says Lindbergh's flight to Paris really "wound his spring." He graduated from Marquette University in 1936, but could not find a job that Summer and decided to go to Law School. In the Spring of his final year, a former Reserve Student Officer (one of about 100 who received flight training 1932-35), by the name of Jim Sliney, came to the campus to show the recruiting movie, "Navy Wings of Gold." Delo says when he saw the F4B fighter planes and the good-looking girls on the Pensacola beaches, he could not wait to get to the place where he could make an application. That turned out to be the Naval Air Reserve Base at Glenview, Illinois, near Chicago. He reported one Saturday morning with 99 others, and he and Tommy Evert were the only two that passed. They took their E Base training there as Seamen 2nd Class. There was no general mess on the base, but the Navy contracted with a farmer about a mile from student quarters to furnish them three meals a day. They marched to and from the farmhouse for each meal, but Delo says it was well worth it, as the food was great. They both soloed and went to Pensacola, but Tommy was one month ahead of Delo in class 129 C. Tommy was later killed in the crash of a B-24 in Europe during the war, but Delo went on to fly thirty years for the Navy and is enjoying his retirement.

Al Dodson, now living in Leawood, Kansas, can't remember when he decided to be an aviator. As a boy, he was reading paperbacks about World War I pilots while other kids were reading Westerns. He went to Central College in Fayette, Missouri, for two years, but was forced to leave because of his father's health and poverty. However, two good friends at Central College, Boone Guyton, and Bart Randolph, were in the first aviation cadet training class (81 C) at Pensacola, and they had a very positive influence on Al. When he read Boone's article in *Redbook* magazine entitled "Wings of Gold," he made up his mind that he would take the naval aviation route in becoming a pilot, if at all possible. Between 1935 and 1939, Al was employed in everything from banks to pipeline companies. He took every opportunity to buy flight time when he had money and to scrounge time with barnstormers when he did not. He flew with Ben Gregory in his old Ford tri-motor plane around Kansas City during that period. Al would pass through the crowds at the airport selling tickets for rides, then he would go to the door and board the passengers, collecting the same tickets he had sold. Once passengers were on board, he would go forward and sit in the co-pilot's seat, even though he didn't know the elevator from the rudder. In 1939, when the Navy lowered educational requirements from four years to two, Al passed his flight physical at Fairfax Airport, soloed before completing his 30-day training period, and went to Pensacola as a cadet in class 133 C.

Harvey Hop, also a cadet in class 133, now lives in Ft. Lauderdale, and owns and operates a Lear Jet charter service out of the airport (Hop-a-Jet, Inc.). He says he has had a natural curiosity about flying since he was a child making balsa models. When he was 10 years old, he used to play "hooky" from school, ride his bike to the grass airport in Holland, Michigan, and watch barnstormers take people for rides. He had his first flight with national hero Clarence Chamberlain when he was 16 years old. He sat in the front cockpit with Chamberlain and was ecstatic during the whole flight. In 1939, he was working in the Wall Street area of New York City for an insurance firm for $18 a week, and he hated it. One day a co-worker brought in a notice from the newspaper indicating that the Navy needed pilots — see the world and receive average pay of $200 a month for four years. Harvey was impressed, just as he had been with the Navy Air Exhibit at the World's Fair, so he decided to do something about it. He made an appointment to take a flight physical at the 90 Church Street office, and he reported thirty minutes early. By the time the exam was scheduled to start there were 18 hopefuls. Harvey and Gerald Rahill were the only two that passed. They took their E Base training at Floyd Bennett, living in the upper bay between the hangars. There were 15 in the class, and all but one soloed. He joined the Air Force and was a Major long before any of the Navy group made Lieutenant Commander. Harvey and cadet Fish were the first in the group to report to Pensacola in 133 C, and although Fish had more than 300 hours pilot time, he "busted-out."

Harry B. Hooper, who now lives in Santa Cruz, California, was a kid who was always interested in flying. He had his first two rides in 1930. He graduated from St. Mary's College in 1939, along with his second cousin, George Dooley. Harry decided to work for six months, but cousin George got into the Marine Aviation Cadet program after graduation and went to Pensacola in class 131 C. George liked the program, and Harry was thinking about changing jobs, so he decided to look into the Navy's cadet program. He successfully passed the flight physical and was told to report to the next "aviation elimination class" at Oakland on 1 March 1940. After he reported, a Navy Lieutenant told Harry that because he was thinking of making a career in the military, and because he had four years of college, he might have a better chance in the Marine Corps. Cousin George liked the Marines, so Harry decided to go that route. He and eight Seamen were in the class, and three washed out. Harry was in class 146 C which had 56 cadets, including 3 Marines. The Marines and 33 others got their wings.

Donald R. Munro, of Birkenfeld, Oregon, says he got curious about flying in the late 1930s, when a gentleman who owed him money ($100) became slow in paying. The gentleman ran a flying service, and Don decided to take the debt out in flying lessons to see how he liked it. The lessons were offered at Pearson Field, Vancouver, Washington, and Don took them until he soloed. He really had the flying "bug" when he returned to his home near Portland, and he learned that the Taylorcraft dealer would give free flight lessons to anyone who purchased a new Taylorcraft. Free lessons continued until students received their private pilot's license, so Don decided to buy a plane. The lessons were given at Swan Island, Portland's commercial airport, and when an airline flight was due, a large red banner was placed in front of the administration building to warn small aircraft to stay clear of the field. The plane Don bought was a side-by-side cabin plane, with a 65 horsepower Lycoming engine; the price was $2,000. At about this time, private pilots were becoming acutely aware of the immense cost of flying, so Don decided to look into the Navy's cadet program. When he reported to a selection board in a downtown Portland hotel, he was asked his age. When Don said he was 26, the Marine Major in charge said he was too old. Don re-read the recruiting folder and noted that applications were invited at the Sand Point Naval Air Station. He inquired and was told to report for a flight physical the following Monday morning. He passed the exam, and subsequently received a letter from the chairman of the selection board saying he was accepted and to report for E Base training at Sand Point in August 1940. Don had no trouble flying two out of three Up-checks, and on 16 October 1940 he was sworn in as an aviation cadet and was on his way to Pensacola. He was commissioned Ensign, USNR, in May 1941, well before the deadline of his 27th birthday.

Jack L. Poteat, of Atlanta, Georgia, became interested in flying in 1935, when Chelly Charles flew an OX-5 into Sam Yancey Field in Marion, N.C. The field was owned by Jack's father, who gave Chelly permission to use it for his barnstorming flights. The pilot sat in the rear seat and carried two passengers in the front seat for $2.50 each. Jack was hooked on flying after that first ride, and he has enjoyed every one of the 40,000 hours pilot time since. He applied for the Navy cadet program in July 1939. He was one of 19 that took the physical in the new Post Office building in Atlanta, and the only one that passed it. Because he had no idea when he would be called for flight training at Opa Locka, Florida, he returned to North Carolina and accepted a teaching job for $98 a month. The Navy called him in November, but he requested a deferment until June 1940, which was granted. He was one of sixteen in his E Base class, and all but one soloed and went to Pensacola. Jack's class was 150 C, which contained that famous cadet Marine Joe Foss from Sioux

Falls, S.D. There were 63 in the class; 45 got wings.

Harry Cocowitch, of Pensacola, was in cadet class 159 C, the class of exceptions. There were 82 in the class and 80 were commissioned in six months' time. The achievements of this class far surpassed those of any other cadet class at Pensacola. Harry had been at the University of Alabama and had begun his flying career as a part of the Civilian Pilot Training (CPT) program in 1938. He was one of 40 students who received flight instruction at Parks Air College in East St. Louis, Missouri. All instruction was given in J-3 Piper Cubs. Students were allowed to solo when instructors thought they were ready (Harry soloed after 8½ hours of dual), and flying continued until students were able to pass flight and written tests for a private-pilot's license. Once the students received their private licenses, 10 of the original 40 were offered advanced training in Myers G-3 planes designed for aerobatics. When cadets in 159 C got to Pensacola they all had at least 40 hours of pilot time, and only 2 "busted-out."

Although many of the motivational stories were quite similar, two more should be included. Art Farwell, of Pensacola, attended a private military school then chose Purdue University because of its outstanding engineering course. During his first year there, he became active in the Purdue glider club. He amassed 500 glider flights, including one world's record at Elmira, N.Y. — 8½ hours soaring in a glider built for three. Most other gliders were built to carry one or two people. He set a National Collegiate record for spot landings (18 inches) at Wayne County airport, in Detroit, in 1936. He also received powered flight instruction and hob-nobbed with luminaries like Amelia Earhart (her plane was sponsored by Purdue and was called the flying laboratory). Art soloed in an unlicensed "Pietenpohl," made by an Indianapolis furniture manufacturer, and powered by a souped-up Model A Ford engine. Art was sworn in as Seaman 2nd at Glenview, Illinois, where the Navy shared the grass field with Curtiss-Wright. He soloed there in August 1937 and reported to Pensacola as a cadet in October, in class 107 C.

The other motivating story that must be included is that of Gordon Egbert of New Orleans. He attended Utah State as an engineering student and in the Summer of 1934 worked as a student engineer in a California CCC camp for $30 a month plus room and board. He had never considered flying, or military service, and was about to graduate after five years of hard study, when he learned about the Navy's cadet program. He passed the flight physical at the Oakland Reserve Base and became a Seaman 2nd Class in July 1937. On his first instructional flight with ex-Reserve Student Officer Harry Sartorris as his instructor, Gordon almost joined the Caterpillar Club (to qualify as a member, one had to "bail-out" of an aircraft and land safely using a parachute). Harry had climbed to 4,000 feet and was demonstrating the stall characteristics of the plane. After about four stalls, he increased the throttle in a dive for the start of an inside loop. It was not until the top of the loop that Gordon realized that he had not fastened his safety belt. He says, "Man did I hang on! That's the last time I ever forgot to fasten my safety belt before a flight." Gordon soloed at Oakland in August 1937 and was assigned to Pensacola cadet class 107 C in November.

Most cadets, except those that came in through the Civilian Pilot Training program, were recruited through one of the Naval Air Reserve Bases, or through Naval District headquarters. All, except a few Marine cadets who went to Pensacola after selection, had to undergo a 30-day training period as Seamen 2nd Class, USNR, at one of the twelve Reserve Air Bases, and the officers in charge did nothing to make life easier for the poor Seamen.

Ninety-five percent of all cadets began their flight training at one of the twelve Naval Air Reserve Bases in the rating of Seaman 2nd Class, USNR, the lowest the Navy had. The flight and living conditions were quite primitive by most standards, and the young college graduates had to have a burning desire to learn how to fly in order to put up with them. The takeoff and landing fields at all twelve bases were grass, and the airplanes were biplanes held together with wire and struts. They had wing skids at the tips of the bottom wing, and a metal skag to keep the tail off the ground. They had no starters and no brakes. There was no communication with a control tower. When a pilot was ready to taxi out for takeoff, he signaled the tower by moving ailerons, elevators, and rudders, and the tower would give him a green light. When he reached takeoff position he received another green light indicating he was cleared for takeoff. There were very strict course rules where the Reserve Base was also used as a commercial airport, as training planes were not allowed to cross the commercial runway in use. If an airline plane was scheduled to land or take off, a training plane in the air would get a continuous red light at least ten minutes before the scheduled time.

The quality of instruction was outstanding, and instructors considered the safety of their students above all else. From the time instructors began preparing students for their first flight, they stressed the danger of the propellers on the ground, and the absolute necessity of keeping flying speed in the air. They noted it was also necessary to land straight ahead into the wind and into the nearest clearing if the engine failed below 500', the lowest altitude at which to consider a turn back to the field. They pointed out that if an untenable situation in the air was imminent, the parachute should be used as the safety of personnel was paramount and equipment was secondary. When in the air above 500', maintaining flying speed and being constantly alert for other aircraft was vital, but the most important factor was knowing surface wind direction and selecting a safe place to land if the engine quit.

Internal communications from cockpit to cockpit were through a gosport (speaking tube), and left a lot to be desired. They were all one way, from instructor to student, and thus students could not talk back. The true word was often "garbled," with mostly profanity coming through. Aware of the poor system, instructors spent much time on the ground with students after each flight, extolling the good points and damning the bad ones. Instructors often cross-checked their students before a solo check. In other words, the tenth hour of instruction was a flight check by another instructor in preparation for the actual solo check. In general, the E Base instructors were not only well qualified, but they were extremely fair and mature in judging students, so that few students were eliminated without due cause.

All twelve Reserve Bases had different living facilities for their Reserve Seamen. I knew those at NRAB St. Louis, Missouri, which also served as the commercial airport. After graduating from Indiana University on 5 June 1939, I reported to the Navy hangar at the St. Louis Airport on Sunday, 15 June, for 30 days of active duty as Seaman 2nd Class, USNR. After being logged in, I was issued a mattress, sheets, pillow, and pillow slip, and was escorted to living quarters in the American Airlines hangar. There were eight cots and eight lockers on the second floor above the airline office spaces; otherwise the area was bare. All toilet and shower facilities were in the Navy hangar about a block away. We received a dollar a day food allowance. There was no mess, nor was there any food facility at the airport, but there was a small resturant across the road, about a mile from the Navy hangar, that did all of the catering for the two airlines that serviced the airport. We ate most our meals there.

American Airlines had one plane going East and one going West each day. Chicago & Southern (now Delta) had one plane from Chicago and one from New Orleans each day. American was using Douglas planes and Chicago & Southern was using Lockheeds. Coffee and donuts was a 10¢; sandwiches were 10¢ and 15¢, and there was usually a choice of plate lunches from 25¢ to 35¢. We didn't get fat, but we did not starve.

We reported in on a week-end, and that Sunday a carrier fighter squadron landed at St. Louis for refueling. The first thing the pilots were looking for was the "head." Thinking they meant the Commanding Officer, we told them he wasn't aboard. That made the fighter pilot's trip, and gave the Skipper, Harry Horney, a laugh on Monday morning.

On Monday we were issued one pair of brown shoes, a cloth helmet, goggles, and several denim shirts and trousers. (I can't remember the Sailor's cap, but we probably had one.) A laundry truck came to the hangar on Tuesdays and Fridays for pickup and delivery, and as I remember, the price was right.

Our training planes were N3Ns manufactured by the Naval Aircraft Factory in Philadelphia. They were more sturdy than the Curtiss Fledglings and the NYs of earlier days, but they were still biplanes held together with wire and struts. Instead of skags they had tail wheels, and instead of pull-through props they had inertial starters. They had brakes, and no bamboo wingtip skids, but they were about as underpowered as their predecessors. We had many ground nomenclature check-outs, and before our first flight we had a session with one of our instructors, who were Bart Randolph and Frank Burchum. The check pilots were Lt. Harry Horney (Commanding Officer) and Al Emerson (Executive Officer). Bart and Frank were newly-made Ensigns out of cadet classes 81 C and 83 C, respectively, and both were about to complete their contracts. Frank was my instructor, and before the first flight he spent about twenty minutes going over the plane and describing what we were going to do in the air. After reviewing the yellow sheet (it showed any discrepancies of the plane on the previous flight), Frank signed it, while I was signing out a parachute from the loft. We went to the airplane; he got in the front cockpit and I got in the back, but before he gave the signal to crank the starter he cautioned me to make a full adjustment to the seat and rudder pedals (he could see I was "short-stuff," 5'6", just barely over the minimum). Then he gave the mechanics the signal to crank the inertial starter, and when it reached the right pitch, they went to the front of the airplane and gave the "all-clear." Frank said, "Contact," and started the engine. After it had warmed up, he waggled all controls and got a green light from the tower. He gave the hand signal to pull the chocks, and we began taxiing to takeoff position. When we reached that, on the grass outboard of the commercial runway in use, Frank waggled the controls and received a green light for takeoff. The only instruments in the rear cockpit were the altimeter and needle-ball (bank and turn indicator), as the airspeed indicator had been blocked out. This was only my second plane ride, but we were not 100' in the air until I knew I had made the right decision! I was doing exactly what I should be doing — learning how to fly and getting paid for it. This made it easy for me to learn, and when it came to my solo check ride with Lieutenant Horney, I passed every test he gave me with flying colors. When he was satisfied, he said, "Take me back to the hangar." When we got back, he stepped out and said, "You got it, Hill, good luck!" I came in a little "hot" for my solo landing, but it was a lot easier with the airspeed indicator. Six of the eight soloed and went to Pensacola. Bucky Walters was in class 129 C, and the other five in 130 C. We reported in September 1939, and three of the five in our class got our wings.

Chapter Two

The Real Test

Although the Navy ordered its first airplanes from Glenn Curtiss in 1911, it did not establish a flight training base until 1914, the year World War I started. The site chosen was Pensacola, Florida, and it was superb as 70 percent of the days were flyable and there was abundant space for expansion. Flight training was accelerated rapidly during the war years and thereafter, but fewer than 4,500 pilots had been designated naval aviators by the time the first cadets arrived in July 1935.

Cadet Walter A. Hibbs was one of the first, and he made one of the longest trips to get there. He came by train from Seattle, Washington, and reported that the trains got slower, smaller, and more expensive as they came East and South. He made it on his allowance with 50¢ to spare, but if he had been going the other way, he would have been far short. A handful of cadets were fortunate enough to own cars, but most of them traveled by rail and had to take a taxi from the station to the main gate.

There were 69 cadets in the first class (81 C). They came from all walks of life and all sections of the country. Some had safely soloed. All were different, but all had one common thread — "a burning and intense desire to be naval aviators." They expected to be in the air that first week and were ready and eager to face the real test of learning how to fly.

At the first muster on their first Monday morning, they received the first of many disappointments. They learned it would be six weeks before they set foot in an airplane, so there was no rush to draw flight gear. The next few days were spent in checking into the various station activities, and one of the most important stops was the dispensary. There they received the most thorough and most critical physical exam to date. Flight Surgeons wanted to "weed-out" cadets who were on the borderline physically, and one of the 69 in the first class was eliminated immediately, leaving 68 in the struggle to survive.

The barracks were well constructed two-storied brick buildings, with 4 wings, each housing up to 75 cadets. Each wing had 12 wash basins, but only 6 mirrors for shaving. There were 6 open showers, 6 urinals, and 6 heads (toilets). The bunks were metal double deckers with good sized mattresses, and each cadet had a spacious locker and some shelving. Each wing had a large screened porch, which was ideal for early morning calisthenics. The entire ground floor of a central building was used as the cadet mess, and meals were served in two shifts. Cadets contributed a $30 per month allowance to the mess. The food, by most standards, was excellent and was served in good style.

Once the check-in sheet was completed, students knew the location of most station activities. For indoor recreation there was the library, movies every night, bowling alleys, and pool or billiards. There were tennis courts, volley-ball courts, and baseball, softball, and football fields available for outdoor sports. There was water in every direction, which provided good fishing, sailing, and swimming, and the Gulf beaches on Santa Rosa Island had the prettiest white sand in the world. It cost $2.00 per month to join the Officer's Club at Mustin Beach where beer was 15¢ a bottle and bar drinks were 20¢. You could have a large evening with a date on Friday or Saturday night for two or three dollars. In the service department, you could get a

Cadet Barracks, Naval Air Station, Pensacola, as they were in the 1930s (top) and as they are in 1983.

haircut for 25¢, a suit pressed for 30¢, and you could get 120 pieces of laundry finished each month for only $3.00 (a pair of socks counted as one piece).

The class was measured for uniforms the first week, and each member was issued one dress blue, two dress white, and four khaki working uniforms. They were also issued one pair of black, one pair of white, and two pair of brown shoes (aviation students and aviators were the only officers in the Navy who could wear brown shoes).

Authorities at Pensacola had been turning out aviators for years before the cadets arrived, and they only had to worry about teaching them how to fly. They did not have to teach them how to be officers and gentlemen, nor did they have to teach them about the Navy, because they were a part of it. For cadets, however, it was a whole new ball game; since they knew little or nothing about the Navy. They needed special attention and a very thorough indoctrination course.

Lt. Commander B. Studley was given the title of Commander of the Aviation Cadet Battalion and was assigned the task of preparing and implementing the Indoctrinal Training Syllabus. He was a "Mustang" (officer promoted from enlisted status), and a logical choice to head this new officer battalion. He knew everything these officers should know before they were sent to the fleet, and he included most of that information in the syllabus. He took a special interest in cadets, shepherded them through the training course, and helped advise them in submitting their choices for duty at the end. He prided himself in knowing the names of every cadet on sight during the first two years of the program. He kept them busy from 5 a.m. until 8 p.m. every working day for 6 weeks. (See Exhibit 1 for syllabus.)

EXHIBIT I

INDOCTRINAL TRAINING SYLLABUS

PERIODS

1. ARTICLES FOR THE GOVERNMENT OF THE NAVY
Reading of Articles 1-4 inclusive — The origin of the Articles — The Mission of the Navy and reason for the establishment of the Articles — Duties of officers and men under the Articles. **1**

2. A COMMISSION
Reading of a commission — The origin of commissions — Early commissions for specific purposes — The standard and uniform commissions of today — The President as the commissioning authority — The significance of a commission — The obligations incurred in acceptance of commissions. **1**

3. A FITNESS REPORT
Reading of excerpts from Article 137 Navy Regulations — The preparation and forwarding of fitness reports — Officer's records — Selection for promotion — Analysis of the qualifications marked on a fitness report. **1**

4. NAVAL POLICY
Reading of U.S. Naval Policy (General Board Publication) — Discussion of features. **1**

5. TYPES OF NAVAL VESSELS
(a) Battleships: early battleships — dreadnoughts — eighteen battleships under the treaty — displacement — armor — main, secondary and AA batteries — torpedoes — magazines — fire control methods and equipment — catapults — power plants and speed — damage control — communications — battleships of foreign navies. **3**
(b) Battle cruisers: characteristics of foreign battle cruisers.
(c) Aircraft carriers: aircraft squadrons — armament, power plant and speed.
(d) Heavy cruisers: displacement — armament — speed.
(e) Light cruisers: displacement — armament — speed.
(f) Destroyers: displacement — armament — speed.
(g) Submarines: displacement — armament — speed — underwater handling.

6. SHIP AND SQUADRON ORGANIZATION **6**
(a) Ship organization: Commanding Officer — Executive officer — heads of departments — division officers, divisions, petty officers and seamen — warrant officers — marines — flag and staff officers.
(b) Squadron organization — squadron commander — executive officer — administrative duties of officers — enlisted personnel — divisions and sections.

7. NAVAL ORGANIZATION **3**
(a) Forces afloat: U.S. Fleet — forces, types, squadrons, divisions — Asiatic Fleet — Special Service Squadron — Naval Transportation Service — flag commands.

(b) Shore Establishment: Navy Department — Offices and bureaus — Naval Districts.

8. FLEET AVIATION — 2
Types of fleet aircraft — carrier squadrons — operation of aircraft from carriers — squadrons on battleships and cruisers — catapults — recovery of planes at sea — patrol squadrons — tenders — aircraft armaments — gunnery practices.

9. QUALIFICATIONS OF NAVAL OFFICERS: — 1
Discussion of subjects covered on examinations for promotions.

10. EXAMINATION ON ABOVE — 1

11. AVIATION TRAINING PROCEDURE — 4
Lectures preparatory to commencement of flight training.

PART II INTRODUCTORY LECTURES — PERIOD

1. GENERAL INTRODUCTORY LECTURES — 2
Fleet construction program and its aviation requirements — Plan of cadet training — Status of cadets — Military conduct and requirements — Confidential matter — Miscellaneous.

2. HISTORY OF NAVAL AVIATION — 1
Beginnings of naval aviation — Pioneer flying — Naval aviation in 1917-1918 — Development of fleet aviation — Development of aircraft carriers.

3. MILITARY DRILL — 1
Purpose of military drill — Inculcation of the habit of unified thinking — Necessity of unified thinking in the Naval Service.

4. TOUR OF STATION — 4
A tour of the Naval Air Station under guidance of an instructor — Witnessing of a catapult shot.

5. AVIATION MEDICINE — 4
A series of four lectures by a flight surgeon.

PART III NAVAL COMMAND AND PROCEDURE

1. MILITARY ETIQUETTE — 1
Salute — how and when to salute — Addressing superiors orally and in writing — Military bearing, wearing of uniforms — Colors — Differentiation between orders and commands — Command channels for cadets — Official correspondence — definitions of secret, confidential, and restricted correspondence — Safety of each — Corresponding Army and Navy ranks.

2. QUARTERDECK, BOAT ETIQUETTE AND STAFF OFFICERS — 1
Quarterdeck, definition and purpose — OOD post of duty — Men of War, boarding and leaving — Small boats, embarking and disembarking, salutes, passing, overhauling.

3. SHIP ORGANIZATION — 1
Chain of command aboard ship — Heads of departments, definition, duties, material and equipage for which each are responsible — Departments aboard, equipage under each, subdivisions of each department.

4. OFFICER OF THE DECK — 1
Definition, duties underway and at anchor, composition of watch underway and at anchor — Log sample entires underway and at anchor — responsibility for log — list of reports made and to whom.

5. DIVISION OFFICER — 1
Definition — duties of, for engineering, deck and communications — responsibility to heads of departments — Description of a Watch, Quarter and Station Bill and responsibility of division officer for — General Drills, purpose and description.

6. JUNIOR OFFICERS, WARRANT OFFICERS AND ENLISTED RATES — 1
Definition — detailed description of instruction received aboard ship — responsibilities — status of aviation cadets — Warrant officers, how appointed and duties — Description of enlisted rates.

7. HONORS AND CEREMONIES — 1
Purpose, description of uniforms, guns, ruffles, flourishes, music, side boys and flags for honors — When honors are rendered — Honors due President, royalty, Secretary of the Navy, and all Army and Navy ranks — Rules for

saluting — National salute — Anniversaries and holidays — full dressing ship — Where national flags and personal flags are flown for each rank.

8. NAVY TITLES AND ACCOUNTS 1
Definitions and purpose of ship and shore titles — Responsibility of supply officer for material and equipage — Detailed description of title B material — Description of submitting surveys — Navy appropriations and accounts — Description of preparing, submitting, and passing Navy budget by Navy Department and Congress.

9. AN ENGINEERING AND GUNNERY YEAR ABOARD SHIP 5
General description of all gunnery and engineering practices aboard ship — Detailed description of short range battle practice — Theory and purpose of gun and torpedo directors — Theory and purpose of a range finder — General description of fire control equipment aboard a battleship — Description of a torpedo practice, a torpedo, destroyer day and night division practice, a depth charge practice — Description of engineering trials — Communication tests — Quarterly and annual inspections, Navy Yard overhaul and drydocking — Fleet maneuvers and purpose.

10. MILITARY LAW 3
Forms of military jursidiction — specifications — precept — cept — rules of evidence — detailed court procedure. Duties of judge advocate, members, counsel and convening authority — Definition and sources of naval law.

11. MOOT COURT 2

12. EXAMINATION 2

PART IV SEAMANSHIP

1. NAVAL PHRASEOLOGY 2
Definition of standard naval terms, phrases, commands and expression.

2. ROPE 2
General lecture on rope manufacture, manila, hemp and wire; designation of lines; common varieties of small stuff; effect of dampness and practice regarding wet lines; uses of all type lines.

3. KNOTS 4
Practical work in tying, square knot, granny knot, double carrick and single carrick bend, bowline, bowline on a bight, single and double blackwall hitch, etc.

4. SPLICES 1
Practical work in splicing — eye, long, short, and back splice.

5. SMALL BOATS 4
Rules of the road; sound signals and their meaning; pulling boat orders; rudiments of sailing; duties of boat officers; duties of boarding officer.

6. GENERAL SHIP CONSTRUCTION 8
Hull construction; armor, ventilation system; drainage system; watertight integrity; nomenclature of decks and compartments; numbering of decks, doors and hatches; protection against underwater explosion; principles of gas and damage control.

7. PRACTICAL SHIP HANDLING 9
Five hour cruise on ship. Each cadet stand officer of the deck watch, act as helmsman, lookout, signal watch. Cadets to assist in mooring and unmooring. A short course of instruction on the guns.

PART V MILITARY DRILL

1. SCHOOL OF THE SOLDIER 8
Colors, right dress, count off, attention, rest, at ease, stand at ease, parade rest, eyes right, facing movements, hand salute, uncover, cover, march, double time, half step, right (left) step, by the flank, to the rear, oblique, route step, change step.

2. SCHOOL OF THE SQUAD (close order only) 8
Squads right (left), about, column movements.

3. SCHOOL OF THE PLATOON (close order only) 8
Right (left) by squads, right (left) front into line, column movements.

4. MANUAL OF ARMS **8**
Order, trail, port, inspection, present, right (left) shoulder, parade rest, stand at ease, rifle salute.

5. SCHOOL OF THE COMPANY (close order only) **6**
Open (close) ranks, inspection routine, column of squads right (left) front into line.

Chief Boatswain Baker, left, the Warrant who moved my car from C.O.'s parking space.

To facilitate handling, each class was divided into left and right wings. Meal hours were different for each wing, and when flying started, one wing flew in the morning and went to ground school in the afternoon, and vice versa, and they alternated this procedure each week.

The "City Fathers" at Pensacola were well aware of how the increasing numbers of cadets would affect the local economy, and they wanted to do as much as possible to entertain them. However, the city was small (35,000) and facilities were limited. The Junior Chamber of Commerce was most active in this area, as it tried to sponsor one dance a month for cadets and worked hard to get as many eligible girls as possible for these events. Local families were good about inviting cadets into their homes for meals, and aboard their yachts for weekend outings. There was one good movie house, and two good restaurants. Carpenters, was *the* "Drive-in" between town and the station, and the San Carlos Hotel had a good bar and juke-box dancing. The Casino on the east side of town had a large dancing area with a juke-box. The few cadets who had cars rarely had to buy gas because they were in great demand. The city was nine miles from the main gate, the bus service was lousy, and most people used taxis in groups.

The most popular place to eat in the area was not even in Florida. It was a farm house across the Alabama state line, named "Bartells," after the German owner. About twenty people could be served at one time, and a new group could be served every hour on Friday and Saturday evenings and Sunday afternoons. Everything was served country style (all you could eat), which included fried chicken with biscuits and gravy, vegetables, a dessert, and homemade wine. If you did not have reservations, you were served homemade wine on the porch while waiting to be seated. If there was a party of more than four or five people, it was best to have reservations, and this was accomplished by designating one pilot flying solo to drop a note at the farm house. From the first classes until long after I arrived on the scene, "Bartells" was a favorite place to have a going-away party for classmates who had "Busted-out." The price of the meal was $1.25, and the wine 10¢ a glass.

Just as the City Fathers were concerned for the welfare of cadets, Navy authorities took great pains to explain to cadets that they were officers and gentlemen and were expected to conduct themselves as such at all times. They were not to embarrass the Navy or fellow officers by unseemly conduct while ashore. They were made painfully aware of this attitude when about six of them in class 82 C were involved in a fracas downtown before they ever started flying. The details were never made public, but these fellows who temporarily forgot they were gentlemen were sent home immediately and never got another chance to fly for the Navy.

One other aspect of civil-naval relations was sexual. For the Navy's part, it required that all cadets leaving the base carry prophylactics, which, naturally, had to be accounted for. So each cadet had to sign for his supply, whether he was going to use them or not. Standard practice in our group was to forge the name of one very popular classmate — who has ever since enjoyed an overblown reputation!

One of the most flagrant violations of officer-like conduct in the air was "flat-hatting" (zooming at low altitude over populated areas, or along the beaches). This was considered very dangerous, and several cadets down through the years failed to get their wings because of such foolish conduct. Any dangerous maneuvers over populated areas by aircraft of the Navy could not be tolerated at any time.

One of the finest things about Navy flight training at Pensacola was the outstanding safety record. The first reporting cadets were told that there had not been a flying fatality there for more than three years. This record was marred, however, on 24 September 1935 when one of four Boeing fighters bombing a target on Santa Rosa Island caught fire at about 2,000 feet. The pilot, an Ensign, who was about to finish the course, tried to "bail out" at about 1,000 feet, but his parachute was afire and it was too late. There had been a crash of

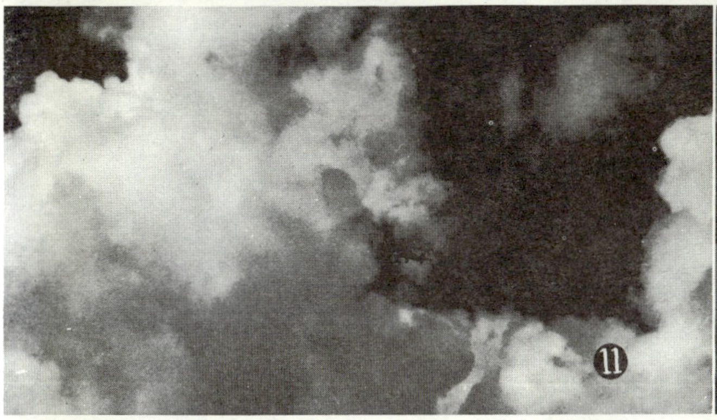

another Ensign and his instructor eight days earlier during a final check in primary landplanes, while making flipper turns over pylons at low altitude. Neither the instructor nor the student was seriously injured, but the plane was demolished.

In both crashes, the emergency procedures were activated, and they were as follows: two vehicles were standing by on radio crash frequencies, as well as ambulances at each field. One of the vehicles was a service-type seaplane with a pilot ready for takeoff during flight hours, and the other was a large tug with a crane, the *Mary Anne*. If the plane observing the crash had radio, he switched to crash frequency and gave the location. The ready plane, the *Mary Anne*, and the ambulances proceeded to the site and gave all possible assistance. If the plane spotting the crash did not have radio, it proceeded to Squadron One and zoomed the tower. This alerted the ready plane that a crash had been observed by the zooming plane. The ready plane took off and followed the reporting plane to the crash site, giving instruction by radio to the *Mary Anne* and the ambulances. The crash procedure was most effective in getting assistance to crash sites promptly and was directly responsible for saving the lives of helpless pilots who had crashed on land or in the water.

The number of weeks spent in each phase of flight training was as follows:

Squadron One — Primary Seaplanes	9 Weeks
Squadron Two — Primary Landplanes	18 Weeks
Squadron Three — Observation Landplanes	9 Weeks

(1) Each Cadet is required to practice Radio Code one hour a day . . . scene in the Squadron 2 radio room. (2) Cadet Timberlake asleep under the watchful eye of Cadet Willy Tepuni. (3) After the hop the Student pilot supervises the refueling procedure. (4) Evening muster closes the day. (5) Volleyball provides the necessary relaxation at Squadron 2. (6) Storing the chute carefully assures the next man a safe ride.

Squadron Four — Service Seaplanes	9 Weeks
Squadron Five — Fighting Planes	7 Weeks

Once cadets started flying, they were thinking about it every waking moment and were discussing how they could improve their techniques. They were talking about it during meals, during the bus rides to the squadrons, on the ground at the squadrons, at ground school, and in the barracks at night. While at the squadrons, if they were not in the air, they were helping their buddies get planes started (it took two people to grind the inertia starters to a sufficient number of revolutions to start the plane once the starter was engaged), or in the case of primary landplanes they were pulling chocks and leading the planes to the point where they could taxi for takeoff. In the case of seaplanes, they helped their buddies get planes started with the same inertia starters before they were taken down the ramps and put into the water.

For a detailed list of the instruction stages in the above phases of flight training, see Exhibit 2.

EXHIBIT 2

FLIGHT TRAINING SYLLABUS

STAGE A "DUAL INSTRUCTION" PRIMARY SEAPLANES

Period	Time	Type Flight	Work Covered
1	5	Dual	Taxiing; takeoffs; straight flying; turns; glides; landings; emergencies.
2	4	Dual	Quarter spirals from 250 feet and half spirals from 500 feet to landing.
3	1	Dual	Entry into spins and recovery. Spins not to be practiced.
4	0.5	Check	Examination on all of above.

STAGE B "PRIMARY SOLO" SEAPLANES

Period	Time	Type Flight	Work Covered
5	5	Solo	Practice periods 1 & 2.
6	1	Dual	Power landings. Spins from skids.
7	5	Solo	Practice periods 1, 2, & 6. Spins not to be practiced.
8	1	Dual	Full spirals. Spins from skids.
9	4	Solo	Add full spirals to previous work. Spins not to be practiced.
10	0.5	Check	Examination on above. Student's work here should be greatly improved over solo check. Takeoffs, landings, and spirals should all be good, and emergencies executed well.

STAGE C "ADVANCED SOLO" SEAPLANES

Period	Time	Type Flight	Work Covered
11	1	Dual	Reverse control turns, spirals & spins.
12	5	Solo	Practice period 11, except spins.
13	1	Dual	Precision landings. Spins from stalls.
14	5	Solo	Practice periods 11 & 13.
15	1	Check	Final check examination on all work.

STAGE "A" DUAL INSTRUCTION

Period	Time	Type Flight	Work Covered
1	7	Dual	Taxiing, takeoffs, climbs, turns, spirals, glides, landings, spins, and emergencies.
2	0.5	Check	Examination of above.

STAGE "B" PRIMARY SOLO 1

Period	Time	Type Flight	Work Covered
3	5	Solo	Practice work of period 1.
4	1	Dual	Review work of period 1; reverse control turns, spirals, spins, and advanced emergencies.
5	4	Solo	Practice work of period 1 & 4.
6	1	Dual	Review work of period 1 & 4.
7	2	Solo	Practice work of period 6.

7

8

9

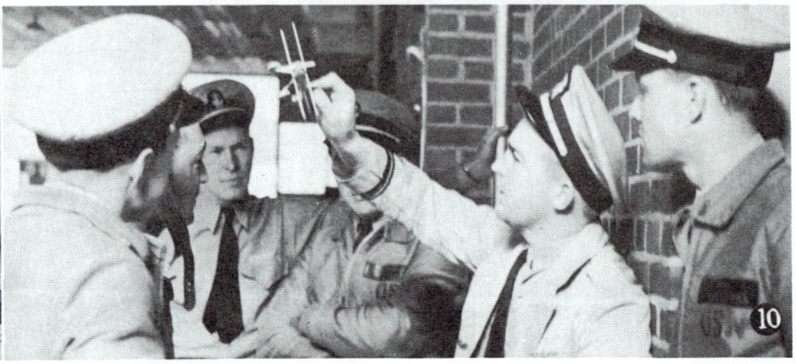
10

11

12

13

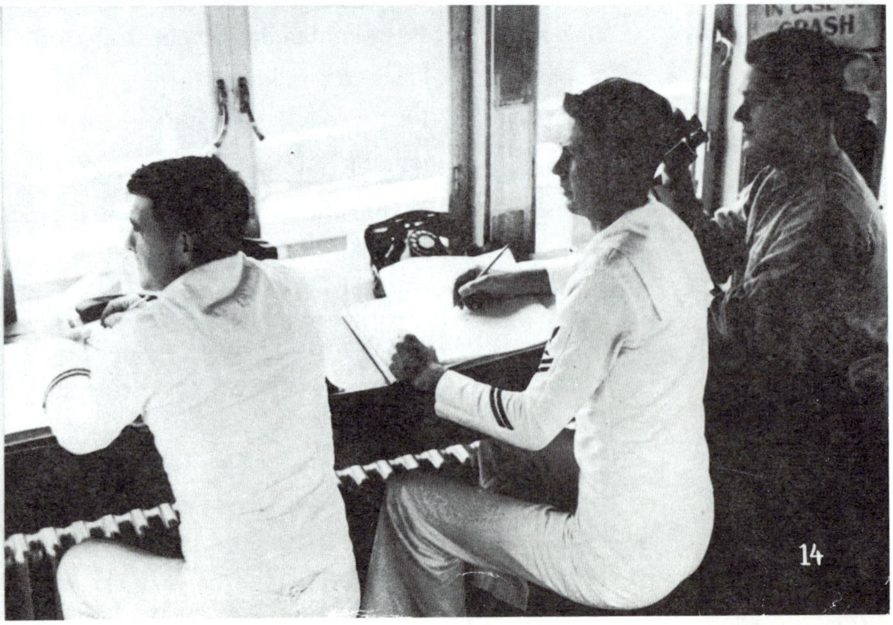
CRASH
14

8	0.5	Check	All work to show great improvement over solo check, and improved technique in emergencies.
			STAGE "C" PRIMARY SOLO 2
9	1	Dual	Precision landings, 180 degrees from 800 feet, 360 degrees from 1,200 feet, precision spins.
10	5	Solo	Practice work of period 9.
11	1	Dual	Review of period 9; intermediate field work; emergency landings, fish tails, slips, precision spins.
12	5	Solo	Practice work of periods 9 & 11.
13	1	Dual	Review of periods 9 & 11.
14	4	Solo	Practice work periods 9 & 11.
15	1	Check	25 hour. Examination of all above.
			STAGE "D" ADVANCED SOLO
16	1	Dual	Splits, loops, wingovers and eights, rolls, precision spins, and review of period 11.
17	5	Solo	Practice work of period 16.
18	1	Dual	Immelman turns, falling leafs, cartwheels, review of period 16.
19	5	Solo	Practice work of periods 16 & 18.
20	1	Dual	Review of periods 16 & 18 plus emergency landings in intermediate fields.
21	3	Solo	Practice work of periods 16, 18, & 20.
22	1	Check	38 hour. Examination of all above.
			STAGE "E" FORMATION
23	2	Dual	Three-plane formation. Simple "V."
24	4	Solo	Practice work of period 23.
25	1	Dual	Three-plane formation; "V" to echelon, position shifts.
26	4	Solo	Practice work of period 25.
			STAGE "F" CROSS-COUNTRY
27	1	Dual	Cross-country instruction; forced landings in emergency fields; poser landings in strange fields.
28	1.5	Solo	Supervisory small field practice.
29	3.0	Solo	Triangular cross-country short flight, no landings. Long cross-country with 1 landing.
			STAGE "G" FINAL
30	1	Solo	Review precision landings and primary work.
31	2	Dual	Final instruction — general review in preparation for final check.
32	1	Solo	Practice work of period 31
33	1	Check	Final examination of syllabus.
			STAGE "H" FRONT SEAT AND NIGHT FLYING
34	1	Solo	Student flying from front seat.
35	0.5	Dual	Night flying instruction.
36	0.5	Solo	Night flying — Instructor to remain with student as safety pilot.
37	1	Check	Night flying.

Opposite page: (7) Over the tower on the south course. (8) The first night hop is an experience that no Student Aviator forgets. (9) The last plane rounds the Administration tower. (10) "Fly it through," just like this. (11) The course is changed. (12) The trainer line as seen from the tower. (13) Mild winters in sunny Florida! . . . but just in case Cadet Frank Swartz wraps up snugly for an altitude hop. (14) Checking the planes as they take off and land on the "world's busiest mat."

Squadron 3 service types (2 O3Us in bottom half and 2 SBCs in upper half).

OBSERVATION PLANE SYLLABUS

STAGE "A"

1	2	Dual	Taxiing, takeoffs, normal turns, spirals, landings, and emergencies.
2	0.5	Check	Examination of above.
3	8	Solo	Practice all primary work, wingovers, eights, flipper turns, spirals, and make three observed and marked landings at Station field after each flight.
	0.5	Check	All of above.

STAGE "B"

4	12	Formation	Three- and nine-plane formation.

STAGE "C"

5	7	Formation with radio	Student pilot acts as radio operator. Practice tactical and procedure signals, simulating service conditions.

STAGE "D"

6	2	Spotting	Student pilot as radio operator. Practice in spotting and procedure signals, simulating service conditions.

STAGE "E"

7	7	Navigation	Student pilot as navigator & radioman.

STAGE "F"

8	6	Cross country	Triangular CC and beacon flying.

STAGE "G"

9	2	Gunnery	IBP (free guns)
10	1	Gunnery	SABP (Camera guns)

ADVANCE SEAPLANE SYLLABUS

STAGE "A" OBSERVATION SEAPLANES

1	1	Dual	Instruction in landings, taxiing, takeoffs, approaches for hooking on.

STAGE "B"

2	1	Dual	2 students practice work period one.

STAGE "C" INSTRUMENTS

3	1	Dual	Instruction in instrument flying, spirals, change of course, etc.
4	7	Dual	Practice work of period 3 with student safety pilot.
5	1	Check	Examination above.

STAGE "D" CATAPULT SHOTS

6	0.5	Dual	4 shots (2 in front seat and 2 in rear)

STAGE "E" TORPEDO PLANES

7	5	Dual	Takeoffs, landings, turns, spirals, glides, emergencies.
8	0.5	Check	Examination on above.
9	0.5	Solo	Taxi practice.
10	4.5	Solo	Practice work of period 7.
11	0.5	Check	Examination on above.

STAGE "F" PATROL PLANES

12	5	Dual	Taxiing, takeoffs, turns, spirals, landings, one-engine turns and spirals, emergencies.

13	0.5	Check	Examination of above.
14	4	Solo	Practice work of period 12.
15	1	Dual	Review of above and instruction.
16	4	Solo	Practice work of period 15.
17	0.5	Check	Examination on above, plus demonstration of towing seaplanes.

STAGE "G" PATROL PLANES NAVIGATION

18	3	Dual	Team of four pilots conduct triangular navigation flights overseas. One as navigator, one radioman, one pilot, and one co-pilot.
19	3	Dual	
20	3	Dual	

STAGE "H" PATROL PLANES

21	2.5	Dual	Two bombing teams, two students on each flight. Each student drops 3 water-filled bombs on each flight. Students alternate as bombers and pilots.
22	2.5	Dual	
23	2.5	Dual	
24	2.5	Dual	

STAGE "I" PATROL PLANES

25	1.5	Dual	Night flying.

FIGHTING PLANE SYLLABUS

STAGE "A" PRIMARY

1	6	Familiarization, landings, takeoffs; last 3 hrs. include loops, spins, aileron rolls, & snap rolls.
2	1.5	Precision landings; carrier landings.
3	2.5	Aerobatics; loops, spins, snap rolls, aileron rolls, Immelman turns, rolls on top of loop.
4	1	Ceiling test — maximum rate of climb to 18,000 feet.

STAGE "B" FORMATION

5	3	Section formation including takeoffs and landings.
6	3	9-plane division formation; all maneuvers.

STAGE "C" TACTICS

7	5	9-plane tactical formation; dives section, slide over, turns.
8	2	Aerobatics.
9	1	Single seater vs single seater (Instruction).
10	1	Section vs single seater.
11	3	Tactical problems; distant rendezvous; attacks.

STAGE "D" NIGHT FLYING

12	2	Night flying familiarization
13	1	Night flying section and division formations instructors leading.
14	1	Night flying, section formation.

STAGE "E" GUNNERY

15	5	Dummy runs IBP (fixed guns)
16	9	Firing runs IBP (fixed guns)
17	6	Dive bombing IBP (bombs)

STAGE "F" NIGHT FLYING

18	4.5	Night cross-country on lighted airways.

STAGE "G" CARRIER LANDINGS

19	5	Carrier landings.

Final flight Squadron 3 is cross country to New Orleans in SBUs. (7) Planes join up at 6500 feet. (8) Off for New Orleans. (9) SBU is a beautiful bird in flight. (10) New Orleans. (11) Pilot and radioman check navigation somewhere over Mississippi. (12) Pilots rehash trip.

Cadets were graded on everything they did at Pensacola on the Navy scale of 2.5 to 4.0 (passing to perfect). They were graded on every phase of the indoctrinal course and every phase of ground school. They were graded on every dual instruction flight and every check flight. If they got a mark below 2.5 it was a down-check, and they were brought before the Squadron Flight Board. This board consisted of the Squadron Commander, the student's instructor, and the check pilot. In most cases the first flight board gave the student three extra hours to practice, and then he had to fly two out of three up-checks with different check pilots. If the student failed to accomplish this, or if he got a second down-check in the same squadron, he was usually assigned a new instructor and given three more extra hours of dual instruction. The extra time in any

Navy built primary seaplane N3N-1 being hoisted from beach in Squadron 1.

Squadron 5 and instrument flight training in SNJs lower left foreground, and "dogfights" in F4B-4 fighters parked in double rows upper right.

squadron could not exceed six hours. If a student got a down-check after six hours extra instruction in a squadron, his case was referred to an Advisory Board, which was ordered to make a report to the Commandant. The Commandant required a report by the Senior Medical Officer to accompany the report of the Advisory Board. Probably less than five percent of cadets received extra time from the Commandant, but the few that received the extra two hours of instruction usually completed the course.

To make a pilot effective in the fleet, he had to know a lot besides how to fly a plane. He had to be able to communicate (send and receive Morse code). He had to know something about the construction of planes and their limitations. He had to know about engines, their operating limitations, and their overhaul. He had to know about gunnery and bombing, and he had to know navigation in order to get to his targets and back. Cadets were introduced to all of these things in what was called ground school, and the instruction time alloted each division was as follows:

Structural Division	95 hrs.
Power Plant Division	115 hrs.
Gunnery Division	60 hrs.
Communications Division	155 hrs.
Photographic Division	10 hrs.
Navigation Division	40 hrs.

After all ground school and all flying was completed, cadets had to pass one final test. We had to send 20 words per minute and receive 25 words per minute in mixed Morse code, all without error. If you did have errors, five words per minute were subtracted from your final score for each one made.

For a detailed list of ground school courses see Exhibit 3.

EXHIBIT III

THE REAL TEST

When students weren't flying, they were going to ground school. This was considered a very important part of pilot training, and required a total of 38 weeks, which included 6 weeks of indoctrination. Courses were broken down into divisions as follows:

STRUCTURAL DIVISION

Course	General Outline	Time Allotted
Aerodynamics	General considerations of airfoils, use of wind tunnels, parasite drag, stability, controllability, maneuverability, the propeller, performance and dynamic loads, principals of design.	20 hrs.
Aircraft Construction	Materials, shop processes, construction of parts; assembly and alignment; inspection and tests; airplane maintenance; airplane accessories, including parachutes.	35 hrs.
Airplane Overhaul Practical Work	The complete process of disassembly, cleaning, repairing, replacing parts where necessary, refinishing, assembly, and alignment is performed. Squadron airplanes are received for overhaul and subsequent to reconditioning and flight tests are returned to respective squadrons.	40 hrs.
	Power Plant Division	
Airplane Engines	General principles; the mechanical system; fuel, lubricating, and cooling systems; fuels and oils; service type engines, and maintenance.	45 hrs.
Test Stand Trouble Shooting Practical Work	Operating and trouble shooting of air-cooled and water-cooled engines of various service types.	30 hrs.
Engine Overhaul Practical Work	Disassembly, cleaning, grinding valves, repair and renewal of defective parts, alignment, test.	30 hrs.
Electricity	Fundamental principles of electricity, stressing applicability to ignition circuits, wiring, and timing of engines.	10 hrs.

GUNNERY DIVISION

Aircraft Gunnery G-1	Aerial deflection, aerial gun sights camera gun, care and cleaning, synchronizer gears, accessories.	6 hrs.

Practical Work	Sighting.	1 hr.
Speed Stripping Practical Work	Nomenclature, disassembly, assembly, and action of Lewis and Browning machine-guns.	5 hrs.
Bombing and Bu-Ord Pamphelts	Theory of dive- and horizontal bombing, description of Mark XI and Mark XV bomb sights, bombing tactics, practical work with bombing training.	21 hrs.
Torpedoes. BuOrd pamphlets	Description, routine unkeep, and tests of Mark VII torpedoes, torpedo tactics.	1 hr.
Range Gunnery Practical Work	Eight practices with stationary and pendulum targets using Lewis and Browning machine-guns and turning mount, pistol shooting, and trap shooting.	25 hrs.
Aerial Smoke Screens	Description of equipment and material for heavy and light aerial smokers, tactics of smoke screen units.	1 hr.
	COMMUNICATIONS DIVISION	
Radio Code Practical Work	Receiving and sending of continental Morse Code, straight and mixed groups Quality 15 wpm code, and 20 wpm press (these requirements were increased 5 wpm)	115 hrs.
Communication Procedure	Formulation, sending and receiving of dispatches by both standard and abbreviated procedure, essentials of communication instructions, U.S. Navy.	10 hrs.
Practical Work	Formulation, sending and receiving of dispatches by both standard and abbreviated procedure, essentials of communication instructions, U.S. Navy.	20 hrs.
Radio Material	Special requirements and characteristics, theory and operation of spotting sets (sending and receiving); care and adjustment, radio telephone sets for aircraft.	5 hrs.
Semaphore	Elements of receiving and sending messages by semaphore.	
	PHOTOGRAPHIC DIVISION	
Naval Photography	Uses of elements of photographic optics; photographic chemistry and material; aerial mapping.	10 hrs.
	NAVIGATION DIVISION	
General Information	History and development of Aviation service, foreign and civil.	5 hrs.
Aerial Navigation	Aerial dead reckoning, piloting, astronomical observations including short methods, navigational instruments, chart and amp reading.	30 hrs.
Aerial Scouting	Methods and doctrine of aerial search contacts, tactical scouting, tracking, patrol and carrier.	5 hrs.

Although everyone realized the importance of ground school, the real guts of the course and by far the most exciting part, was flying. We all knew it was dangerous even to be near an airplane, and far more so to be in charge of it on the ground, or in the air. Safety of personnel was the primary consideration, and this was stressed from the moment you talked with your instructor about the first flight. Once you were in the air, you were taught to plan what to do when the engine quit. For the first 200 feet there was no choice — you maintained flying speed and control and landed into the wind. Survival in this case was about 100 percent, but to lose control and spin, chances were zero.

In the first four cadet years (1935-39), all students began flight training in Squadron One, primary seaplanes. Training was done inside the limits of Pensacola Bay, which was 25 miles in length. It was ideally suited for seaplane work, as Santa Rosa Island provided a barrier that cut off the ground swells from the Gulf of Mexico. The planes were NY-2s, on pontoons with wing-tip floats, powered by 250 horsepower Wright Whirlwind engines. Engine failure in seaplanes was not as critical as in landplanes, because you had open water for landing. Before flying began, planes were rolled on dollies and launched down concrete ramps for the first flight each day. At the end of each day's flying, planes were picked up by cranes mounted on railway cars, placed on dollies, and rolled back into the hangars. One of the most amazing sights on Pensacola Bay each morning was to see forty to fifty seaplanes milling around awaiting the takeoff signal (two orange balls were hoisted on the tower). It was a real "rat-race" when everyone tried to take off at the same time, because the last one in the air was a "Pelican." Some of the wise instructors counseled their students to let the "eager-beavers" take off, because it was much better

6

7

8

9

10

11

12

13

Everyone's favorite stunt plane in Squadron 2 at Corry Field.

Some early birds soloed in NY's which were built by Consolidated Aircraft Company.

Opposite page: Cadets Rosbert and Sherwood(129-C) made P3M deadstick landing in East Bay when engines quit. Plane had to be towed to the Pensacola Bay bridge, and eased under the draw bridge where it was picked up and towed by NAS crashboat.

An O2U from the cruiser ***Raleigh*** *flying over the Fleet. Floatplanes such as this were used by the cadets at Pensacola for advanced training. (Vought)*

to be a live pelican, than a dead cadet.

Once students completed all work in Squadron One and successfully passed all check flights, they started all over again with seven hours of dual flight instruction in primary landplanes in Squadron Two. There was much more traffic in the squadron at Corry Field, but it was far more orderly and safer than the traffic on the bay in seaplanes. The course rules there were very strict, and one of the surest ways to "bust-out" was to land, or take off, on the wrong course.

After solo check, there were precision spins, split Ss, fishtails, small-field emergencies, and landings in a 100-foot circle. Then aerobatics, and the students were instructed in completing loops, wingovers, split-S turns, snap rolls, Immelman turns, falling leafs, cart-wheels, and figure-eight flipper turns over pylons at low altitude. This was the real fun part of flying, as you had to take charge and make the airplane perform. Once you passed the stunt check, the pressure was off, because you could pass the flight requirements for a commercial transport pilot's license, and you had it made. Before you finished the squadron you had three-plane formation work, and then came the final check. If you received an up-check, the check pilot gave you a precision 100-foot circle shot. If you hit the circle, you brought the plane to a full stop, exchanged cockpits, and flew the instructor back to Corry Field in the front cockpit.

When you got to Squadron Three, you were expected to be a pretty smooth flier. Instructors allowed only two hours to get used to the heavier service-type planes before giving some precision landings and a solo check. The headquarters for the squadron was Station field, and O2Us (Vought Corsairs) with 450 horsepower Wasp engines were used exclusively for the first ten hours. After eight hours and a final check, you were sent to SU scouting planes with radio. You had one week of radio while formation flying, then had camera gunnery and two days of live machine-gun fire at "windsock" targets towed by other airplanes. You received one week of navigation practice, and then flew a final cross-country formation flight to the Army Air Corps Base at Maxwell Field, Montgomery, Alabama. By this time, you had finished ground school, and all you had to worry about was completing Squadrons Four and Five. The pressure was gone.

Before you went to Squadron Four, there was a short transition period in service type seaplanes in Squadron One that provided one of the greatest thrills in the entire course. After some familiarization in the Vought Corsairs on floats, you had four catapult shots. You were accelerated from a standstill to 62 miles per hour in just 52 feet. There were no instructors, we students did it all. On the first shot, you were a student observer in the

Outboard spotting of F4B-4 to save flight deck spotting space on early carrier.

rear cockpit, and on the second shot, you were the same in the front cockpit. On the third shot, you were the student instructor in the front cockpit, and on the fourth shot, you were the instructor safety pilot. It sounds more complicated than it really was. About the only thing you had to remember as the pilot in control was to keep your head firmly back against the headrest just before the shot. Otherwise your head would be snapped back against the rest, knocking you out, and leaving control of the plane with the observer.

After the catapult shots you went to Squadron Four and the first instruction in T4Ms. These were Hornet-powered Martin Torpedo planes with three cockpits, and they were controlled by a wheel like the steering wheel of a car, rather than the conventional control stick that had been in planes up to that time. Instruction in these heavy T3M torpedo planes assisted in the transition to the larger and heavier patrol flying-boats, which were twin-engined P3M planes equipped for horizontal bombing. We used the bay east of the Pensacola bridge for our operating area here, while Squadron One used the bay west of the bridge. We had familiarization and check flights, but most of our flying was with groups of our contemporaries in capacities as pilots, co-pilots, navigators, bombadiers, and radio operators. I spent almost three months in the squadron, and flew 137 hours, but my training was interrupted by an apendectomy which grounded me for six weeks.

The final seven weeks were spent at Corry Field in Squadron Five. Emphasis was on carrier landings, aerobatics, and instrument flying. There was also a ceiling test, with maximum-rate of climb to 18,000 feet, gunnery, dive bombing, and night formation over airways. When the cadets finished Squadron Five, they had to pass the radios code test, were given their wings, designated naval aviators, and received orders to the Fleet.

The first year of cadet training was a great success. (July 1935-July 1936). There were eight cadet classes (81 through 88 C), and there was not a flying fatality among them. The quality of instruction was high, and the end product that went to the Fleet was outstanding.

Top: Martin T4M torpedo bomber. (Naval Aviation Museum)
Above: For others this T3M-1 built by Martin was a new torpedo plane in the fleet.
Right: T3M-2 (Martin-Marietta)

Stearman N2S-4, a popular training plane. (Naval Aviation Museum)

Precepts were set, tradition was established, and the standards those early birds set for this new rank in the Navy were first rate.

Walter Hibbs turned out to be the top man in the first class. His leadership ability was recognized by Lt. Commander Studley before indoctrination classes began, and he was made a Wing Leader and Battalion Commander immediately. He had been commander of the banner NROTC battalion at the University of Washington. He liked the Navy and knew a lot about it, but his first love was medicine. He wanted to follow in the footsteps of his father and become a doctor, but funds were not available to go to medical school and he decided it would be fun to try aviation. He did not like Pensacola or its climate, as it was too hot and humid. Despite this, he loved flying and breezed through all courses and flight checks. He was the first of 35 classmates sent to the Fleet. He even kept track of where they were sent to active duty. (See Exhibit 4.)

Only 2 of the 35 elected to join the Regular Navy and make a career of it. John Joseph Lynch was the first cadet selected as Rear Admiral, retired in that rank, and in 1983 lived in the Washington, D.C., area. Vincent Francis Casey retired as a Captain to Cape Coral, Florida. Eight in the group were killed in plane crashes before the end of World War II. Nine went with Pan American Airways, and three went with TWA. Six were in civil service with the CAA or the FAA. Three were test pilots, one each with Douglas, Curtiss-Wright, and Vought. Three were surveyed from aviation for medical reasons, and one became a flight instructor at Glendale, California. After he finished his Fleet duty, Walter Hibbs went back home, married his sweetheart Frances, and flew with PanAm as a Captain until he retired in April 1972 to Carmel, California.

Down through the years, cadets had the same problems. Time was limited, they had to budget it, or they ran out. The working weekday started at 5 a.m., when duty personnel from the Officer of the Day's office came through the barracks and turned on the lights, and "lights out" was at 10 p.m., Sundays through Thursdays. This left seven hours for sleep. In the morning there was always a crude wake-up call and out to calisthenics, followed by open showers. The latter also tended to generate ribald humorous remarks of the sort that young men, and even older ones, indulge in in such communal situations. But in the twenty years I was in the Navy, and in all of the time since, I have yet to hear about an ex-Navy or Marine cadet who was ever involved in homosexuality. One problem at night was to "pipe down" those who snored loudly. They invariable wound up on the top bunks so that their buddy below could kick them when the snoring got out of control (I always had the top bunk).

EXHIBIT 4

THE FIRST AVIATION CADET CLASS, 81 C

Name	Hometown	College	E Base	Duty
J. E. Abraham	Uniontown, PA	Duke	Philadelphia	VS-41
H. R. Bailey	Forest Hills, NY	Trinity	Floyd Bennett	VBM-2
A. C. Barnard	Randolph, MA	Boston	Squantum	VP-8
J. C. Barrett	Birmingham, AL	API	Opa Locka	VF-3
A. H. Bohne	Los Angeles, CA	UCLA	Long Beach Air	1-FMF
(Commissioned USMC 8 July 1936)				
R. A. Bowen	Minneapolis, MN	Minnesota	Minneapolis	VMB-2
T. B. Bradbury	Louisville, KY	Princeton	Floyd Bennett	VF-6
R. O. Buschmann	Indianapolis, IN	Purdue	Oakland	VP-8
V. F. Casey	Los Angeles, CA	USC	Long Beach	VF-4
F. S. Chamberlain	Corvalis, OR	Oregon St.	Seattle	VMS-2
L. Gogliani	E. Boston, MA	Tufts	Squantum	VS-3
G. S. Felt	Sparta, MI	Michigan	Grosse Isle	VS-42
B. A. Foulds	Jasper, MN	Minnesota	Minneapolis	VMS-2
R. S. Garrison	Long Beach, CA	Wash. St.	Long Beach	VB-2
W. E. Gise	Hollywood, CA	UCLA	Long Beach Air	1-FMF
(Commissioned USMC 8 July 1936)				
A. S. Gowen	Atlanta, GA	GA. Tech.	Opa Locka	VB-1
B. T. Guyton	E. St. Louis, MO	Central	St. Louis	VB-2
M. G. Haines	Beach Grove, IN	Purdue	Great Lakes	1-FMF
(Commissioned 8 July 1936 USMC)				
W. A. Hibbs	Sequim, WA	U of Wash.	Seattle	VT-2
R. J. Jensen	Holliston, MA	Boston	Squantum	VB-3
W. H. Jones	Lake Helen, FL	U of Florida	Pensacola	VF-5
A. N. Kelly	San Diego, CA	San Diego St.	Long Beach	VF-5
K. A. King	Brooklyn, NY	St. John's	Floyd Bennett	A-1-FMF
(Commissioned USMC 8 July 1936)				
G. H. Knott	Seattle, WA	U of Wash.	Seattle Air	1-FMF
(Commissioned USMC 8 July 1936)				
J. J. Lynch	Dorchester, MA	Boston	Squantum	VB-3
F. R. Payne	Corning, NY	U of Arizona	Floyd Bennett	1-FMF
(Commissioned USMC 8 July 1936)				
G. R. Randolph	St. Louis, MO	Central	St. Louis	VF-4
G. D. Rohrbaugh	York, PA	Penn St.	Philadelphia	VP-10
T. M. Roscoe	Two Harbors, MN	Minnesota	Minneapolis	VP-10
K. F. Schader	Seattle, WA	U of Wash.	Seattle	VB-4
M. D. Smith	Minneapolis, MN	Minnesota	Minneapolis	VB-1
J. R. Vensel	Butler, PA	Carnegie Tech.	Anacostia	VB-2
P. B. Withers	St. Johnsbury, VT	Norwich U	Squantum Air	1-FMF
(Commissioned USMC 8 July 1936)				
H. W. Younkman	Montclair, NJ	Brooklyn Poly	Floyd Bennett	VB-2
D. K. Yost	Princeton, NJ	Princeton	Pensacola Air	1-FMF
(Commissioned USMC 8 July 1936)				

This exhibit shows the wide distribution of cadet hometowns, colleges, and E Bases, in the first class. Later classes had a smaller percentage of Marines, fewer assignments to duty in carrier squadrons, and larger numbers to patrol squadrons.

Despite crowded conditions and some inconveniences, cadets learned to make the best of them and to cooperate with each other for the good of all. We all knew the course was no "piece-of-cake," and that we had to give it our very best shot to finish it. If we did not have a nickname when we arrived, we were soon given one by our shipmates. It may have come from our accent, the way we dressed, the way we walked or marched, the way we ate or drank, the way we flew, or from distinctive mannerisms that made us different.

The Navy's number one fighter plane F4B-4 from the time first cadets started training in 1935 until they finished their four year Navy contract. (Boeing)

Instructors as well as students had nicknames. Two were called "Jumpin' Joes" (Clifton and Eastman). They were nervous types, always bouncing up and down with their heads on a swivel, and sometimes riding the controls. One check pilot named Wier was referred to as, "Wier today and gone tomorrow," since he gave downs to about two of every three pilots checked.

We were all very close to each other, and it was a sad day when one of the group "busted-out." To soften the blow for those who were so unfortunate, most every class collected a dollar each payday as "bust-out" insurance for those who did not make it. We all had a great amount of respect for each other.

How about the instructors? They were all different, but they were all human. Some were aggressive, loud, and profane, while others were passive, quiet, and firm. Some could fly better than others, and some could teach better than others, but they were all dedicated and most of them were fair. They were all aware of the constant hazards of flying, and the safety of their students was paramount. In Squadrons One and Two, instructors had extreme difficulty in getting verbal instructions across. There was a "gosport" (speaking tube) that ran from the front cockpit to the rear cockpit and was plugged into our helmets. Communication was one way, as we could not talk back, and it was usually "garbled" (anything but clear). Unless instructions were spoken clearly and distinctly, about the only words that came on strong were profane, because the cockpits were open and the engines made a lot of noise. I'm sure many students could not pass their flight checks because they simply did not get the "word." Positive hand signals were developed to assure control responsibility of planes at all times.

One of the more famous stories about hand signals and "gosports" in Squadron Two involved an instructor who prided himself on his ability to determine when his student was safe for solo. If this was determined during a check ride, he would reach down and loosen the set screw that held the control stick, lift the stick from its socket, hold it up before the student, and throw it over the side. He would then pat himself on the head and point to the student, meaning the student had control. Then the instructor would put in a spare stick and ride back to Corry with his arms on each side of the cockpit. Everything was fine until the instructor threw his stick over one day and reached down for the spare. The mechanic responsible for putting in the spare had inadvertently put it in the rear cockpit. The instructor was horrified when he looked back at his student in time to see him holding up *his* stick and throw it over the side. The instructor was so upset that he promptly went over the side. The student had seen the spare stick, reached down, and put it in the socket, tightened the set screw, and safely brought the plane back to the field. No more charades for that instructor!

Standard operating procedure (SOP) has always dictated that pilot safety was paramount and that aircraft should be abandoned if there was any question about the safety of flight. In 1980, even with excellent communications that are in training planes today, an instructor at Pensacola "bailed out," and his student brought the plane back safely. On 29 September 1980, Marine 2nd Lt. Steve Hoenie was about to complete his last dual flight with instructor Marine Capt. Dean Lucas before his solo check. Hoenie had control at 2800 feet altitude when a turkey buzzard came crashing through the canopy, knocking out Hoenie and blinding Lucas. Lucas determined the plane was about to get out of control. He called twice for Hoenie to bail out, and when he got no answer, he abandoned the plane. A few seconds later Hoenie regained consciousness, got the plane under control, and brought it back to base where he landed safely. He was one of the lucky few who managed to solo before his check flight.

There is another story about a student in Squadron Three who changed his check flight from a down-to an up-check. The student was making a long power approach to Station Field when his check pilot cut the engine. The student correctly determined that he could not safely reach the field, so he headed the plane into the wind to make a water landing. However, he failed to retract his landing gear, and this was a "no-no" for service type landplanes because when the landing gear touched down on the water, the plane would flip over on its back. The instructor took control, pointed out the error, retracted the landing gear, and was making a final approach to the field when the student waggled the wings. He got the instructor's attention, pointed to the gear, and saved an airplane and the check pilot's rear. The student turned a sure down into an up-check for the record.

Fraternity brother, Don "POP" Korn, stepping into Link trainer for instrument training in Squadron 5.

E-Base classmate, Bucky Walters, checks wind sock before entering cockpit of F4B-4 Squadron 5. Bucky landed torpedo plane at Kaneohe, Dec. 7, 1941 at night, among dozers and cars without a scratch.

Whatever else may have been said about instructors, they were a dedicated group of pilots who had the safety of their students in mind at all times. There were misunderstandings about check flights, and some simulated emergencies turned into actual ones. Some instructors were killed while trying to teach their skills, and more than one was killed by a propeller. With up to 100 props turning over simultaneously, it was impossible to hear anything. You could not safely have your head locked in any position on the ground, or in the air.

You could safely say that check pilots were demanding, but fair. Above all else, not one of them wished to give a student an up-check when he did not deserve it. Check pilots were not about to be responsible for turning out a product that might be a menace to himself or others.

Almost everything that could happen to instructors and students in training planes, did happen. The first cadets had not been flying two months in 1935, when Will Rogers and Wiley Post were killed in Alaska in Post's plane. Students and instructors alike analyzed the fatal error Post made when he tried to get back to the field after his engine failed. He did not have enough altitude to maintain flying speed and control, and the plane spun into the ground. Instructors stressed to

Patrol Squadron 10 about to land at French Frigate Shoals (500 mi. from Midway) during advanced base operations in 1935 when first cadets began flight training at Pensacola.

students that the safest procedure in event of power loss shortly after takeoff was to maintain flying speed and land straight ahead into the wind. Chances of survival in this case were almost 100 percent, but they were about zero if the plane stalled and spun.

Cadet C. O. Rogers, class 83 C, reports that G. L. Pentz, also in class 83, was the first cadet to join the "Caterpillar Club." To qualify for this exclusive club, one had to jump from a plane and land safely with a parachute. Pentz bailed out during a night flight in Squadron Three and was temporarily knocked out by the flashlight that was required to be worn around his neck on night flights. However, he regained consciousness in plenty of time to pull the ripcord and land safely. Rogers also reported that cadet C. D. Shaffer in 82 C was the first cadet fatality when he spun in during his final flight in Squadron Five in September 1936. Two more cadets in 86 C were killed in October that year, but the over-all safety record was outstanding. In 1935, Squadron Two received the Schiff trophy for safety when it flew 18,699 hours without injury to personnel.

Night formation flying and low altitude flipper turns over the pylon course proved to be the most hazardous parts of the training course. D. M. (Doc) Burns in 97 C was one of the pilots in a nine-plane night formation flight of three sections. There was an instructor in the lead plane of each section, and students in the other six. Students were practicing sliding over to make echelon for left and right formation turns. During one such maneuver, Cadet Brennen was crossing over from left to right behind Cadet Bates in the middle section to form right echelon for a left turn. During the crossover Brennen's propeller began chewing up Bate's tail assembly and both planes went out of control. Bates jumped out of his plane without any problem, but Brennen had trouble getting out, his chute ripped before it opened, and his ankle became entangled in a shroud line, so he came down head first. Probably fewer than a handful of pilots have ever been in such a predicament and lived to tell the tale, but Brennen was lucky that night when what was left of his chute caught in the top of a pine tree and left him dangling a few feet from the ground. But it was not long after this incident that the flame went out of Brennen's burning desire to fly, and he abandoned the air for good.

There were several occasions at night when fog rolled in from the Gulf while planes were flying night formation cross-country flights out of Corry Field. On one such occasion in 1938, nine Boeing fighter planes were in formation returning from Mobile. When the flight returned to Corry, the fog had rolled in, so that none of the planes landed there that night. However, Charlie Harbert, class 105 C, decided to make a couple of passes at it. Looking straight down from 2,000 feet, he could see the lights of the field, but in letting down he entered the fog at 800 feet and let down to 200 feet twice without seeing a thing. He finally decided that he had better get out of the situation, so he climbed to 5,000 feet heading away from the city and decided to bailout. The Boeing fighter was not an easy plane to get out of, so Charlie devised his own ejection system. He cut the throttle, put his foot on the stick, and catapulted himself right out of the plane. After his chute opened, the fog made visibility zero, and he could not tell when he was getting close to landing, but he made a soft one in the water. It was up to his chin and moving, and he found that if he moved cross-current in one direction it became more shallow. He decided he was in the Escambia River, made it to the river bank, and walked until he came to a highway bridge. When he got there, Lieutenant Gabbert, the station Duty Officer was there to pick him up. Gabbert was Charlie's first instructor in Squadron One, and the one that told him to let the "eager-beavers" take off, because it was better to be a live pelican than a dead cadet. When he picked Charlie up, he said, "Harbert, the smartest thing you ever did was to bailout tonight."

Cadet flying fatalities ranged from zero the first year (1935-36) to twelve in 1940-41, but there were more students in training each succeeding July. Aside from night formation flying, more fatalities occurred on the pylon courses than in any other phase of training. The course had to be negotiated at an altitude of 500 feet or less, and loss of control at that height was fatal. One of our instructors, J. E. Dalquist, who was also the football coach, was killed at the pylons while giving a student a twenty-hour flight check. The student lived, but he was seriously injured and never flew again.

Aircraft collisions accounted for some fatalities; and there was no hope of getting out of a low-altitude crash alive. I was almost a statistic one day while making precision landings to a 100-foot circle. On the last instructional flight before the twenty-hour check, instructors got out of the plane on the first landing and observed student landings from the edge of the circle. I had made five landings in the circle and was intent on making the final one perfect, when I observed the usual three or four instructors stop their wagering and start running away from the circle. I looked back over my shoulder and saw the biggest propeller I had ever seen about to chew up my tail. I pushed the stick forward and gave it full throttle to avoid a certain fatal collision between cadet Harry Ginn's plane and mine. I made one more pass at the circle to land and pick up Monk Russell, but I was not very close to the center of it. When Monk got in the plane he said, "I thought I was

*The Boeing F4B-4 fighter, this one from the carrier **Saratoga,** 22 September 1931, was every pilot's dream. (Navy Department)*

going to have to walk back to Corry Field with Harry Ginn's instructor. I was sure you guys had bought the farm.'' This incident shook me up so much that I ''busted'' my twenty-hour flight check and had to fly two out of three ''up'' checks before continuing the course. I was assigned a different flight instructor by the name of Severson, who was a Marine Captain and a great pilot. He was also a tough task master and an outstanding instructor who gave me the confidence, the courage, and the determination to become an aviator. The close call while making precision landings to a 100-foot circle taught me a valuable lesson as a student. A pilot should never concentrate so much on any one thing that he forgets to look around. Keep your head on a swivel, and never let it get locked in any one position, in the air or on the ground, when you are near an airplane.

Commander of the aviation cadet battalion, Lt. Commander Studley, worked out a system of demerits that were to be awarded cadets who were operating below acceptable standards. Some cadets had a propensity for picking up these little ''goodies,'' and Walter C. (Mike) Michaels was one of these. He got off to a very bad start in Squadron One when he left his parachute in a seaplane at the end of flying one Wednesday afternoon. The planes had been put in the hangars, and the Duty Officer spent more than two hours looking for the missing chute. Mike was awarded 100 demerits, and the penalty for receiving more than five demerits a week was the privilege of marching around the barracks with a rifle, one hour for each demerit. Despite his many weekends of marching, Mike was still marching off hours several days after he finished the course. Captain William ''Bull'' Halsey, Commandant after he got his wings,and later a famous World War II admiral, pinned on Mike's wings after he had marched the last hour. Although he was a soft touch for demerits, Mike reports things were not always bad. One night after payday he was in a five-card poker game with Greg ''Pappy'' Boyington (later aFlying Tiger and then C.O. of the famous Black Sheep Squadron), ''Red'' Fitzgerald, and four others. After the draw and limit of raises by Mike and Greg, they showed their hands. Greg had a full house with Kings over Queens, but Mike had four Aces for the first time in his life and won the pot. Mike was one of those rare birds who almost had his wings before he learned to drive a car. He had lived in New York City all of his life, his parents did not own a car, and he did not learn to drive until he bought one a few weeks before he went to the Fleet.

I was one of those lucky birds, who escaped marching off demerits. But one day in Squadron Four I parked in the Commanding Officer's parking space. It was a great spot and allowed me to make my flight on time, which saved ten to twenty demerits, but I did not realize it was the ''Skipper's'' space. I did have the good sense to leave the keys in the car, so that it could be easily moved. When I returned from the flight, there was an urgent message on the bulletin board for Cadet Hill to report to the Commanding Officer. When I reported to him, he tossed my car keys to me and said, ''Chief Boatswain Baker gave these to me after he moved your car from my parking space. If your car had been there when I arrived, I would have ''busted'' you. I think you owe Mr. Baker a large favor, so go down there and see what you can do for him.'' After thanking Mr. Baker profusely, I asked him what I could do. He said, ''You're in charge of officer parking; if you spot an unauthorized car, have it towed away.'' I spent many hours policing the officer parking area, but that was much better than marching around the barracks with a rifle.

Little has been said about cadets and romance at Pensacola, but the city was small and there were just so many eligible girls available. Some of those were not attractive enough to get cadets excited, and many of those available had local boy friends. The girls knew that cadets had agreed not to get married for four years, and they were more likely to get serious about one who was about ready to be washed out. However, there were many Pensacola lasses who became Mrs. Cadets. Love was stronger than regulations, and many decided to take their chances. If their marriage was discovered and they were released from their cadet contracts, they would have no problem in getting a job as an airline pilot. The marriage ban of the late thirties would certainly pose no problem today, as a couple would simply strike up a ''love-in'' arrangement and they might not ever consider tying the knot. To ease the shortage of women at dances, it was the custom at Pensacola to allow cadets to cutin on their shipmates. This made the most popular girls more popular, and made more ''wallflowers'' available. The cadets with automobiles had the most flexibility in the romance department. If they did not like what they found in Pensacola, they could go to Mobile, which had more than twice as many eligible girls. Beyond Mobile there was an exclusive girl's school (Gulf Park College), which had several hundred pretty girls. After I purchased a car, I learned more about this flexibility and girls and dated some in Mobile and Gulf Park. I was rather serious about Mary Keene at Gulf Park, and stopped by her home in Kansas City for a large date, on the way to Hawaii. By the time I finished Pensacola, the marriage ban had been cut to two years, and I was determined to wait that long before assuming the additional responsibilities of marriage. Many of us had only a passing interest in girls at Pensacola, as quite a few had love interests back home. Whatever the time limit was, most of us abided by it.

About six weeks before we finished the course, we were allowed to submit three choices for a duty station. There were three types of duty. One was carrier duty, which comprised fighter, dive-bomber, and torpedo squadrons. The second type of duty was in observation, or scouting planes, aboard cruisers and battleships. The third type was in long-range patrol planes. The second and third types were seaplanes operating from water, whereas the first type operated from car-

The N3N-1 "Yellow Peril" as a landplane (above) and the N3N-3 version as a seaplane were the mainstay of the instructional squadrons at Pensacola in the latter 1930s. (Naval Aviation Museum)

riers and were landplane squadrons. In 1936, when the first cadets were ordered to duty, there were only three carriers in operation (*Lexington, Saratoga,* and *Ranger*), and each carried about 100 pilots in 4 squadrons. Battleships and cruisers had 3 to 4 planes each, with about 6 pilots. Patrol squadrons had 12 planes, and about 50 pilots each. It was considerate of the Navy to allow cadets to submit choices, but they had to be assigned where needed. Most were disappointed with assignments, and one I know, Carl Rinehart, Class 121 C, wrote out his resignation when he received his orders to a patrol squadron in Panama. Carl's cadet brother, a year ahead of him at Pensacola, was in a fighter squadron on the *Lexington,* and Carl's first three choices were all on that ship. When he presented his resignation to the Superintendent of Aviation Training, the Commander promptly tore it in half and explained that the Navy was not going to accept it. More than $40,000 had been spent on his training, and the Navy was not about to let him go fly with Pan American or any other airline. Carl got the message and went to New Orleans to catch a "banana boat" to Panama, obeying his orders just like any other cadet. Several other cadets submitted resignations when disappointed with orders, but I do not know of a single instance where one was accepted. The Navy was proud of their cadets; they had a large investment in them and did not want to lose them for any reason.

One of the final tests required before getting our wings and carrying out our orders was to pass the Morse code sending and receiving test. If you had not practiced this procedure, you were likely to be a little "rusty" and make mistakes. You had to send at least 20 words per minute and receive 25 in mixed code (words with numbers, not press). I completed everything except the code test, and then invited my sister, Marjorie, down from Indiana to help me drive straight home, so I would not lose valuable leave. Would you believe that it took me three full days to pass the test, and it cost me two extra nights in a hotel for Marjorie. Finally on 2 August 1940, I passed it, received my wings, was designated Naval Aviator #6547, and was commissioned Ensign, USNR. We drove straight to Indiana in 16 hours without incident.

My orders were to report to Patrol Squadron 23, at Pearl Harbor, Hawaii, TH. I had steamship passage booked on the *Matsonia,* which was scheduled to leave Los Angeles on 25 August. Paul Hart, Lincoln, Nebraska, and Neal Lytle, Cleveland, Ohio, were new Ensigns scheduled to sail on the same ship. Neal had a new Pontiac, and I had a new Plymouth, so we decided we would meet Paul in Kansas City, Missouri, and drive cross-country together. My good friend, Mary "Keenie" Keene, from Gulf Park College, lived there and had agreed to have dates for Neal and Paul and to meet us at the Muehlebach Hotel at 4 p.m. on 20 August. Everyone was on time and we had twelve hours of sheer joy. It was daylight when we got to the hotel, and noon when we checked out, but we made Albuquerque by midnight (650 miles). We reached Las Vegas by midnight the second night, checked into a motel on the outskirts, and decided to go downtown for a beer before the "joints" closed. When we entered the first place, we ordered a beer and asked what time it closed. The bartender said, "We opened in the Summer of 1933 and have been open ever since, so I guess it will be some time before we close." After a couple of beers there, we decided to make our investigations separately and meet back at the motel at 6 p.m. to get some sleep before crossing the desert at night on the way to Los Angeles (there were no car air conditioners in those days). When we regrouped at the motel at 6 p.m., we checked our losses (about $150), set the alarm for 3 a.m., and turned in for some much-needed sleep. We arrived in Los Angeles about 9 a.m., 24 August, and had until noon the next day to get our cars aboard the *Matsonia*. The Navy paid for shipping our cars as well as for our ship passage to Honolulu. The price of the passenger ticket was $125 for double occupancy in a stateroom, and this included five wonderful days of entertainment and outstanding meals.

Besides Paul, Neal, and myself, there were two other new Ensign pilots out of Pensacola, all headed for Honolulu on the *Matsonia*. We donned our dress white uniforms every evening for dinner, which made a hit with the ladies. All five pilots were ready for anything that officers and gentlemen could be caught doing until the wee hours of the morning. We put the ship to bed every night, and all agreed that the five-day passage was the finest trip we had ever made. It was sad to think that it was coming to an end, but it was fun to see the Hawaiians diving for coins and to see the large crowd on the dock waiting with leis to greet the new arrivals. Representatives from each squadron at Pearl Harbor were there to greet us and take us to the BOQ (Bachelor Officers Quarters) at Ford Island with the Fleet. My good friend and classmate at E Base and Pensacola, Bob Corlett, met me and took good care of me. He had a vested interest in my arrival, because I was to relieve him of his duties as the squadron "Yard-Bird."

PART II
FLYING FOR THE FLEET

Chapter Three
Peacetime Operations

Patrol Squadron Twenty-three was one of three PBY squadrons based at the Naval Air Station on Ford Island, Hawaii. The island was about 1½ miles long and ½-mile wide and was surrounded by a deep water channel that would accommodate the Fleet's largest ships. Quarters were provided for about 50 bachelor officers, squadron commanders, and senior officers of the air station.

When I reported to the squadron in August 1940 there were 50 pilots and about 250 enlisted personnel. Fifteen of the pilots were Naval Academy graduates, and the ranking officers in the squadron. Twenty were reserve officers who had all been aviation cadets, and fifteen were Naval Aviation Pilots (outstanding enlisted personnel trained as pilots). All pilots received basically the same flight training at Pensacola. All squadrons and the Air Station enjoyed a tropical schedule of working hours, 7 a.m. to 1 p.m. (0700 to 1300 in the Navy), Monday through Friday. Working six hours straight beginning at 0700 was more productive than coming to work at 0800, taking an hour for lunch at 1200, and trying to work four hours in the hot afternoons. Transportation was provided between the Bachelor Officer's Quarters and work areas.

I lived in Bachelor Officer's Quarters on Ford Island, Hawaii, with about 40 pilots. Each had an adequately furnished room, but we used shower, basin, and head facilities jointly (one facility on each floor). The dining room and kitchen would accommodate about 50 people, and everyone was required to join the mess. We elected a new mess officer each month to plan the meals and handle finances. We ate well, with many fresh fruits and vegetables from the islands, and our mess bill was rarely more than $25 per month. The most inconvenient part of living on the island was the time and preparation involved in getting ashore and back again. Our cars were parked in the Navy Yard at Pearl Harbor, and to get to them we had to take a bus from quarters to the island boat landing, then take a boat to the Navy Yard. The procedure was reversed when we returned to park our cars, with boats running at night only on the hour until midnight, except on Sunday mornings when the last boat left for the island at 2 a.m.

My first day in the squadron was the last Friday in August 1940, and I spent most of it learning about the duties and responsibilities of a "Yard Bird" (junior officer pilot in rank). The day began with the delivery of the Leading Chief Petty Officer's muster report to the Executive Officer, and it ended once the entire hangar area was policed (cleaned) and inspected and all watches posted. Duties in between included all of those not specificially assigned. When the "Yard Bird" was not involved in some constructive task, he was often dispatched to another squadron, or to the overhaul and repair shop, to get a non-existent tool or engine part. My two predecessors who had held this dubious title were Jack Coley and Bob Corlett, Pensacola classmates, who endured it for about 30 days each. Fortunately, I was relieved by George Poulos, Class 132, in about three weeks.

Before I had completed my first month in the squadron, I was awakened one morning at 2 a.m. and told, along with Jack Coley, that we were going on an emergency flight mission. We reported to the squadron and learned that we were to fly with Jack Collett, Flight Operations Officer, in one plane, while "Smoke" Strean and "Sully" Kauffman in another plane were to follow with a Flight Surgeon aboard. The fleet oiler *Ramapo's* 1st Class Pharmacist had an acute attack of appendicitis and the oiler was 950 miles from Honolulu (more than four days at flank speed), so the Commander-in-Chief of the Pacific Fleet decided it was necessary to dispatch a surgeon to the oiler to perform an appendectomy that would save the sailor's life.

Each plane was loaded with 1,400 gallons of aviation gas, which would keep us in the air for more than 20 hours. By 5 a.m. on the morning of 28 September 1940, we had been launched into the ship channel that surrounded Ford Island and were ready for takeoff. There was very little wind, and we went halfway around the

The PBY went through a barely perceptible evolution during its decade of operational service, the most notable change being visible in the fin and rudder, which by Pearl Harbor had been extended upward into the more familiar wartime shape.

island before we were airborne. Planes and pilots were unaccustomed to taking off with such heavy loads, especially in the complete darkness that prevailed an hour before daylight. Once we were airborne, however, we immediately began the momentous task of navigation that would be required to locate a moving ship 900 miles at sea. We did not know what type of weather we would find at the rendezvous site, but we did know that our navigation had to be exact in case visibility was restricted. If we failed to locate the *Ramapo* and safely get the surgeon aboard, our mission was a complete "bust." In the lead plane, Jack Coley and I divided navigational duties under supervision of Jack Collett, while "Sully" Kauffman and "Willie" Howe did the same in the following plane under supervision of "Smoke" Strean. We flew at 1,000 feet altitude within a mile of each other, and we were cross-checking navigational information every 10 minutes. The weather was exceptionally good considering the distance we covered, and visibility was always 5 miles or better. When we reached a point estimated to be 50 miles from target, we flew abreast of each other at 5-mile intervals, and estimated visibility to be 10 miles. We spotted the ship 5 minutes sooner than we had estimated, and it was directly ahead of our two planes. The surface wind was 15-18 knots and the seas were running 6-8 feet. "Smoke" landed his plane without incident, and the surgeon was taken from the plane to the oiler in a whale boat. Once the surgeon was safely aboard, "Smoke" began the first of two aborted takeoffs, as huge swells and heavy seas made it impossible to get airborne. On the second attempt, a wing float was damaged, and "Smoke" decided any more takeoff runs would be too dangerous, so he told Jack Collett to return to base without him while he arranged a tow from the *Ramapo*. We proceeded to Pearl Harbor and landed three minutes ahead of our estimated time of arrival; we had been in the air exactly 16 hours. Nine days later the *Ramapo,* "Smoke" Strean and his crew, and a totally wrecked PBY arrived in Pearl Harbor. It was then we learned that our mission had been a success. The appendectomy saved the sailor's life, but one PBY had to be surveyed (written off), and a new plane was ordered to bring the squadron up to full strength.

Four days before this mercy mission, while on an instrument familiarization flight with Eddie Harshutz, one engine failed and we were forced to land in Kaneohe Bay. We anchored the plane, went ashore in a rubber life raft, and called for the squadron to send mechanics to repair the engine. They arrived within an hour, installed a new set of spark plugs, and had the engine running by late afternoon, and we flew back to Pearl Harbor before dark.

On 7 November 1940 while flying with Phil

Bankhardt on a high-altitude (12,000 feet) 12-plane formation horizontal bombing run during Fleet maneuvers, we had engine failure and were forced to land on 8- to 10-foot waves in the open sea off the island of Lanai. In the earlier versions of the PBY, engine failure always meant a forced landing, because the propellers could not be feathered. It was about 11 a.m. when we were forced down, and the Commanding Officer of the squadron wired Fleet headquarters our position. The Commander-in-Chief of the Pacific Fleet dispatched the aircraft-tender USS *Wright* to proceed to that position and recover our downed aircraft. In the meantime, the native Hawaiians on Lanai had observed our forced landing and came out to the plane in their catamarans to see if anyone was injured. They also brought us plenty of fresh pineapple and papaya. We thanked them profusely for the good food, and for their concern for our welfare, but we told them we were sure that our shipmates had wired for help and that we would be picked up before dark. It was almost dusk when the *Wright* appeared on the horizon, and it was completely dark when the aircraft was taken aboard. We were back in Pearl Harbor the next morning, but the plane had to be off-loaded on a barge and taken to the squadron area where a new engine was installed.

I had now been in the squadron two months and had been co-pilot in two forced landings. But those were the last I experienced due to engine failure. By the end of 1940, I had about 140 hours of pilot time with the squadron, and I had the greatest respect for the PBY. It was slow, but safe, and could do about anything and go any place where there was water. It was the aviation "workhorse" of the fleet.

Long before I got to the squadron, it was called upon to make one of the most historic flights ever made until that time. It was 2 July 1937, and at 0700 muster the word was passed that Amelia Earhart was down, or missing, at the time of her scheduled arrival at Howland Island, and that the squadron might be called upon to send a plane to Howland. When flight preparations began, it involved Aviation Cadet Page W. Smith, Class 86 C, who was to be the co-pilot on the flight. Here is his story:

> Lieutenant W. W. Harvey was chosen to command the flight, with Lieutenant William Crane as Navigator, and Lieutenant (JG) Gene Lytle and myself as co-pilots. Two flight engineers and two radiomen were chosen to round out the crew. The search plane was to depart Pearl Harbor at 1900 (7 p.m.) July 2, 1937, and fly to a point 300 miles Northeast of Howland Island, arriving about daylight July 3rd. The orders read: "Conduct a search from there until a landfall is made on Howland, then make an expanding square search of the area until dusk. Make an open sea landing in the lee of the Island, and tie up astern of U.S. Coastguard Cutter *Itasca* for the night." Then on the morning of July 4th, the search would be continued and further plans would be made. The flight was airborne on time at 1900, and all went well for the first seven hours. At midnight we were cruising at 13,000 feet when we encountered the leading edge of the inter-tropical front, and began picking up ice on the propellers. This was in the days before airborne radar and pressurized cabins, and it was impossible to miss the build-ups because of the dark overcast conditions. We were flying at about maximum altitude, since we had no oxygen and could not get above the build-ups. From the time we were three hours out of Pearl Harbor our radiomen were calling the lost and overdue aircraft continuously, and requesting Amelia to send her carrier wave three times if she could hear our signal. I was listening to one such request before we entered the front, and there was a three signal answer, but who it was we will never know. As we got deeper into the front, we had severe turbulence, hail, lightning, and icing. Also, we were getting bad sea condition reports and forecast from the *Itasca* at Howland. Finally, after eleven hours conditions were so bad, that Lt. Harvey made the decision to return to Pearl Harbor. It was a difficult decision, but I believe it was correct, because to continue would have jeopardized the aircraft and the crew, and we would have been ineffective as a search vehicle operating in such weather and sea conditions. After we cleared the front, the return to Pearl was routine. We did pass over several Navy surface vessels that were steaming toward Howland, including the carrier *Lexington*. We landed at 1910, 24 hours and ten minutes after take-off, and we had four hours fuel remaining out of the 1,500 gallons we had at departure. We were met by newsmen and everyone told his story. I was quite tired since I had been in the co-pilot's seat for 22½ hours, but before I could get away to my bed in the BOQ, Captain Kenneth Whiting, the Ford Island Commanding Officer, asked me to sail with his daughter, Edith, in a Starboat sailing race from Pearl Harbor to Diamond Head and back. As an AvCad, it was impossible to decline the invitation of a four striper, so I went sailing even though I was dog tired.

Page completed his Cadet service in December 1939 and flew for Pan American for 30 years, accumulating 26,000 hours pilot time.

On 3 July 1937 the Commander-in-Chief Pacific Fleet dispatched the carrier *Lexington* and two destroyers to Howland Island to search for Amelia. Many of the pilots and crewmen were on leave and liberty for the July 4th holiday, but a total of 62 pilots from 6 squadrons, and an appropriate number of flight crewmen to man 62 aircraft, left for Howland. The planes searched areas surrounding Howland for six days, covering 151,556 square miles at an altitude of 200 feet or less, all without an accident. However, there was no sign of Amelia or the wreckage of her plane after six days of intensive search, so the ships returned to Pearl Harbor. There was no record of how many pilots in the search were Cadets, but in talking with some of these involved, I believe there were more than 30.

AvCads were involved in many other types of peacetime operations, such as the dedication of the San Francisco Bay Bridge in 1936 and the dedication of the Golden Gate Bridge in 1937. They competed in the Cleveland Air Races every year between 1936 and the war years, and they participated in the dedication of the New York World's Fair in 1939. Many cadets were pilots on the historic flight in January 1937 when Commander "Shorty" McDade and his 12 PBY-1s flew 2,600 miles from San Diego to Honolulu in 21 hours and 40 minutes. This flight demonstrated the feasibility of moving large numbers of aircraft great distances in a short period of time.

Most of the pilot training in patrol squadrons based at Pearl Harbor in 1940 and 1941 centered around preparations for Wing Tactics in November and Fleet Tactics in March. Squadrons would compete for high scores during these maneuvers in high-altitude bombing on moving targets in daytime and at night. They also competed for scores while making live ammunition gunnery runs on towed targets, and while making drops of water-filled torpedoes and depth charges. Advanced base operations were conducted at Johnson Island and Midway Island when detachments would operate from seaplane tenders, sometimes for as long as a week. This was great training for the wartime operations we faced later in the South Pacific. Another part of our training included air-raid drills, and some of these were conducted on weekends, involving carrier planes ashore as well as patrol planes. Although we considered an air raid on Pearl Harbor a very remote possibility, we were very much alert for sabotage. Each squadron had armed security guards around aircraft and hangars 24 hours a day.

Our routine at Pearl Harbor came to a halt in May 1941 as our squadron was one of those chosen to move our base of operations to the new Naval Air Station at Kaneohe Bay, on the windward side of the island. Our squadron number was changed from 23 to 11, and we completed our move in June. The patrol squadron strength in Hawaii was doubled at this time, with three squadrons at Pearl Harbor and three at Kaneohe. The new facilities were great. We had a new BOQ, with an Officer's Club across the street, and the operating area in the bay was ideal for seaplane [flying-boat] operations. It was also great to be able to get in your car and go where you wanted, without having to get a bus and a boat to get to the car. Also, we had a golf course on our side of the island at Kailua, and we could be at the first tee in fifteen minutes. When we were at Ford Island, it took us an hour to get to the first tee at the Army's Schofield course. The only thing I missed at Kaneohe was the good poker games we had when the carrier pilots came in to Ford Island. I was the unofficial gambling host with the responsibility of furnishing the chips and the poker and bridge cards. When one carrier was in, we always had two poker games with seven players at each table, as well as a bridge game or two. One of the poker games was a 5¢ and 10¢ cent game, with a 5¢ cent bet any time and 10¢ cents on the last card. The other game was 25¢ and $1 with the same betting rules.

Not long after we had completed our move to Kaneohe, the word filtered down to Reserve pilots in Hawaii that President Roosevelt had signed an Executive Order authorizing Reserves to resign their commissions and join the American Volunteer Group (AVG) in China. Richard Aldworth headed all

Ensigns Joe Hill and Willie Howe enjoying the Surf Club sand at Waikiki Beach, Hawaii, in the carefree peacetime days of 1941.

recruiting, which for the Marines was done at Quantico, Virginia. For the Navy, it was done at Pensacola, Norfolk, San Diego, and Jacksonville. For the Army, it was done at the following fields: Bolling, Selfridge, McDill, March, Mitchell, Langley, Hamilton, Eglin, Craig, Maxwell, Barksdale, and Randolph. Recruiters were looking for volunteers, 23-28 years old, who were unmarried and had experience in fighter-type aircraft. Everything about this operation was top secret, but the rumor was that the group would be flying the Army's latest (P-40) fighter. They would be fighting Japanese pilots in China, and they would get $500 for each confirmed Japanese plane shot down. The base pay was $250 to $750 per month depending on responsibility in the group, and volunteers were to get 30 days leave with pay, as well as all traveling expenses to China. There was also a $30 per month ration allowance. The first contingent of pilots to leave the United States met at the Mark Hopkins Hotel in San Francisco on 7 July 1941. They departed 10 July 1941 on the Dutch ship *Jaegre Fontaine* and arrived in Rangoon, Burma, on 28 July

Early model PBYs in echelon to port. (Naval Aviation Museum)

1941. Although no recruiting was done in Hawaii, the rumor was that Richard Aldworth was about 15 pilots short of the 150 authorized, and he was coming to Hawaii to fill the quota. He was scheduled to arrive in August, and at least six of us at Kaneohe were looking forward to an interview. My BOQ roommate, Willie Howe, and I were two of the pilots in VP-11 that were to be interviewed. However, one week before Aldworth arrived, Army and Navy Commanders issued a bulletin stating that resignations would not be accepted from Army, Navy, or Marine Reserve pilots until further notice. That ended our anticipation about being ''Flying Tigers,'' and we were bitterly disappointed, because we were sure that we had many friends among those Navy and Marine pilots who had volunteered.

Under the President's Executive order, the Chinese-American Manufacturing Co. (CAMCO) was established, and 100 new P-40 fighter planes were shipped to that company in China. Retired Army Capt. Claire Lee Chennault was chosen to command the group. The recruiting goal called for 150 pilots, 23 to 28 years old, with three years of fighter experience. The actual count was about 20 pilots short, and only a dozen met the fighter plane experience qualifications. As late as March 1942, 18 pilots remained unqualified for combat, and several had been killed in bombing and strafing raids.

Chennault knew he had a tough assignment. Once pilots began reporting, he could see that most lacked fighter plane experience, and that they were going to be difficult to discipline on the ground. This made him determined to demand discipline in the air. He required every pilot to be given 72 hours of lectures and 60 hours of special instruction before he set foot in a P-40. He had made detailed studies of Japanese flight-training manuals and pilot-training tactics. He knew the strengths and weaknesses of their Zero fighters as well as he knew the good and bad features of the P-40s. The Zeros were faster and more maneuverable in a dogfight, but the P-40s were faster in a dive, they had more effective fire power, and they had protective armor plate for pilots.

The volunteers called themselves ''Flying Tigers,'' and the air discipline Chennault instilled in them paid off handsomely. During the first six months of their operations, only four Tigers were shot down, but the group accounted for 60 Japanese planes at $500 each. After World War II officially began on 7 December

1941, Tigers got credit for confirmed Japanese planes destroyed on the ground. Although few P-40s were lost in the air, many became casualties during the rigorous training period before they could get into combat. Few spare parts came with the original 100 planes, so crashed planes had to be cannibalized in order to keep the others flying. Even though the 130 plane mechanics did an outstanding job, fewer than 40 planes were fit for combat after the first six months.

Frank Schwartz, a Pensacola classmate in 130 C, was one of those unfortunate pilots who did not see much combat. He was seriously injured during a bombing and strafing attack on the operating field, and was later taken to a hospital in India, where he died. Another friend, Jack Newkirk, mustered us every morning for calisthenics in Wing 3 during the first few months I was at Pensacola. Jack was an outstanding leader of the group, accounting for many Japanese planes, but he was one of the four shot down during the first six months. Another friend in Wing 3, Robert "Buster" Keeton, was with Jack the day he was shot down. Other cadets I knew at Pensacola who became "Flying Tigers" were Tex Hill 121 C; Noel Bacon, Charles Older, and Joe Rosbert, all in 129 C; Kenny Jernstedt in 131 C; Bob Layher and Peter Wright in 132 C; and Dick Rossi in 134 C. All of these gentlemen except Joe Rosbert are still living today and have been involved in everything from Air Force and Navy careers, to judge, to airline captains, and entrepreneurs.

To sum up the activity of the AVGs, I think it can be said that Chennault and his Tigers did more with less than any other aviation group in history. For a detailed account of their activity, see Chennault's book, *The Way of a Fighter,* and Robert Lee Scott's, *Flying Tiger: Chennault of China*. Also, Greg "Pappy" Boyington, Class 88 C, gives a good account of his activity with the group in his book, *Baa Baa Blacksheep*.

While Flying Tigers were shooting down Japanese aircraft in China, other U.S. volunteers on the other side of the world were making their contributions to our peacetime operations in the war effort. About the time President Roosevelt's Executive Order was creating the Chinese-American Manufacturing Co. (CAMCO), the Navy Department was calling for volunteer pilots from the three Patrol Squadrons in Coco Solo, Panama Canal Zone. The volunteers knew only that the duty was so secret that they could not tell their wives anything about it, including how long they would be gone, because they really did not know. Seventeen pilots, mostly cadets, including Carl Rinehart, went to Washington, D.C., to find out what they had volunteered for. On 5 April 1941, they were assigned to the Naval Operations Department and received secret briefings about their mission. They were given passports titled "Assistant Naval Attache, and Assistant Naval Attache for Air, American Embassy, London, England." On 25 April they went to Elizabeth City, N.C., for outfitting, and on the 28th they flew to Bermuda where they were to pick up new PBY-5 aircraft with skeleton RAF crews. They tested their planes on the first two days of May, rested 3 May, and took off for a trans-Atlantic flight on the 4th. The flights observed complete radio silence and were told to ignore all incoming radio signals because they were probably of German origin. On Carl's PBY-5 (AH 555), the flight lasted 23.5 hours, all dead reckoning, including a 6-hour snow storm at night with zero visibility. They landed at Greenock, Scotland, on 5 May, but AH 555 did not get into service, as it was sunk by a German bomb that night.

The new PBY-5s were a part of our Lend-Lease program to Britain, and the pilots were assigned to the British Coastal Command. They were to fly with the RAF as "Observers," to help their pilots familiarize themselves with the new Lend-Lease planes. Carl was assigned to No. 210 Sqdn., Coastal Command, out of Oban, Scotland, and flew with 1st Lt. Percy Hatfield and his crew of six, which included a radar operator. Carl and the other U.S. volunteers were well aware that the U.S. was not at war, and that if they were captured in U.S. Navy uniforms they would be treated as spies and shot. On 24 May Hatfield and crew picked up a new PBY-5 with special equipment at Helensborough. They flew it back to Oban, where they flight-tested it, and were told to stand-by for a special mission. At 12:15 p.m. on the 26th, Carl took off with Hatfield's crew to search for and locate the German battleship *Bismarck,* which had sunk HMS *Hood* and damaged HMS *Prince of Wales* two days earlier. U.S. pilots "Tuck" Smith and "Jimmy" Johnson had located the *Bismarck* on 25 May, but could not continue tracking it until relieved and had to return to base because of a fuel shortage. Carl's plane flew to the last reported position of the *Bismarck* and started an expanding search. After 18 hours of looking, on the morning of the 27th they came out of a cloud at 4,000 feet altitude and were directly over the *Bismarck*, which was ablaze with anti-aircraft fire. Carl's plane sent a contact report and waited just out of gun range until the British Navy moved in with ships and planes to sink the "unsinkable" battleship. When Carl and Hatfield's crew returned to Oban, they had been in the air 26.2 hours and were welcomed as heroes, but nothing was officially said about U.S. Navy pilots even though the three planes that contacted the *Bismarck* all had U.S. Navy co-pilots. They were not officially there!

Although many cadets were involved in the British war effort and flew practically every type of aircraft they used, they could not be decorated or recognized for their bravery. Carl and most of the other cadets in his group returned to the U.S. before 7 December 1941. In Carl's case he flew in a B-24 which took off from Prestwick and was supposed to land in Montreal, Canada, in 14 hours. The plane encountered headwinds and rough weather conditions, and finally landed at Gander Lake, Newfoundland, more than four hours overdue and 1,000 miles from Montreal.

Back to our situation at Kaneohe, we heard "scuttlebutt" (Navy rumor) that our squadron was to be the first to fly to San Diego for new airplanes. By the end of the first week in September 1941 we learned that the scuttlebutt was true, and we began getting our airplanes in first-class condition. We also learned that six of our planes were to be flown to Jacksonville, and six to Corpus Christi. This was very exciting news to me, and I wanted to be a co-pilot on one of the planes going to Jacksonville, as my girlfriend was in Gulfport and my parents were in Indiana. I had completed the required two years as a bachelor, had been corresponding with "Keenie" for more than a year, and was seriously considering marriage. However, in my letter breaking the news about our trip to Jacksonville, I indicated only that we expected to arrive on 6 October and would be there for two days. I also indicated that my parents would be there, and that I wanted her to meet them. I decided that if she did not come to meet us, that would be the end of our romance.

From the middle of September 1941 until we were beginning our trans-Pac to San Diego, pilots and crews were busy getting planes ready for the trip. Mechanics were busy performing engine checks, and changing engines when required, and pilots were making flight tests, as well as full load takeoffs. Takeoff was scheduled for 3 October. Between 15 September and 2 October I had fifteen flights as pilot and co-pilot which averaged about four hours each. I had a total of 800 hours pilot time and had qualified as a First Pilot (one who could take a PBY on limited missions with another First Pilot), and I expected to be qualified as a Patrol Plane Commander (pilot who could take a PBY on any type of mission) before the end of the year. Cadets who had reported to the squadron before 1940 were now all Patrol Plane Commanders (PPCs). Willie Howe, Jack Coley, Bob Corlett, and I were all First Pilots and flew many limited mission flights together. The number of "Regular" officers in the squadron had been greatly reduced when we moved to Kaneohe, and they were all replaced by "Reserve" cadets. The "Skipper" was LCdr. L. W. "Slim" Johnson, one of the finest, and his Executive Officer was Francis "Snooks" Jones, a Lieutenant. The other Regular Lieutenants who were PPCs were William "Bull" Schroeder and "Red" Rice. Regular Lts. (jgs) Joe Cobb and Jimmy Vaughn were PPCs. The other six PPCs who were to pilot their planes back to San Diego were Reserve Ensigns Bankhardt,

Seen from above the starboard quarter, this PBY cruising down the California coast near San Diego just before World War II clearly shows the new fin and rudder, the distinctive parasol wing, retractable wing-tip floats, and the blister waist-gun positions. (General Dynamics)

Blunt, Heyer, Johnson, Kauffman, and Pennell. Most of the planes in the flight of twelve had at least two PPCs, a First Pilot, a Co-pilot Navigator, and a Naval Aviation Pilot (enlisted pilot). Each plane also had a Chief Machinist's Mate with two mechanics and two radiomen. The planes were topped off before launching with a maximum load of 1,500 gallons of 125 octane gas. The plan was to takeoff at three-minute intervals at about 1100 Kaneohe time (1330 in San Diego) on 3 October and land in San Diego 20 to 21 hours later. The planes were to climb to 9,000 feet, with the lead planes throttling back to slow cruise until the last planes were in loose formation. The Skipper did not want to arrive in San Diego before 9 a.m., because there was usually a morning fog at this time of the year. This prompted the 11 a.m. takeoff time.

On the morning of 3 October 1941, all twelve VP-11 planes were launched on time, and takeoffs began at exactly 1100. I was the co-pilot for Hank Blunt, and our plane was one of those to be delivered to Jacksonville. It was a beautiful morning, with 15 to 20 knots of surface wind, and all planes were off on time. We were plane number nine in the takeoff slot, and we could see all eight planes ahead of us before we lost sight of the islands. The flight was uneventful, and we flew directly over the destroyers stationed at 500-mile intervals to assist us in navigation. We had 10 hours of night flying, with 4 hours on instruments in broken clouds. By 1000 we closed to right echelon formation and could see Pt. Loma. The Skipper touched down at 1030, and the last plane at 1045. We were all a little tired, but it was great to see the U.S. mainland again, and we were looking forward with anticipation to what the next few days might bring.

The first thing we learned was that the group flying to Jacksonville would not leave San Diego until the morning of the 8th, then fly to Corpus Christi, and arrive in Jacksonville on the 9th. I called my parents, then Keenie, about the delay. I suggested they contact the Duty Officer of the Training Command concerning our arrival. My parents said they would be there with Sister Marjorie, Aunt Rae, and Grandma Fox. Keenie was not sure she could make it.

Our six planes for Jacksonville took off at dawn on the 8th and landed in a "Northeaster" at Corpus Christi. The winds were 35 to 45 knots and the waves outside the breakers that protected the launching and recovery ramps were 6 to 8 feet. All six planes managed to get inside the breakers and tie-up to buoys, but we had to keep the engines running at taxi speed until the winds calmed at about 10 p.m. and the beaching crews could pull us up the launching ramps. The next morning we were launched at daybreak, and were winging our way to Jacksonville before 7 a.m. We arrived seven hours later at 3 p.m., losing another hour to time zones. My family was waiting for me outside the Duty Officer's office, when I arrived about 4 p.m., but there was no sign of Keenie. I was free to be with the family until 8 a.m. on the 11th, when we had to begin our return to San Diego by Naval Air Transport R2D2 [USN DC-2]. The family was disappointed in not getting to meet Keenie, but when I told them the romance was over, we settled down to the catching-up we had to do about events of the last 14 months. We had a grand time, but it was much too short.

At 8 a.m. on the 11th we took off in the R2D transport, flew to Pensacola, refueled, and flew to Dallas where we remained overnight. The next day we flew to El Paso, refueled, and flew to NAS North Island at Coronado. We remained in Cornado for two weeks, flight testing our new airplanes a total of fifteen hours. During this period, I decided to call the High School sweetheart in Indiana to see how she was getting along. Her name was Jerry Ringham, and I found she was well on her way to becoming a doctor, as a 2nd year medical student at Louisville University. We talked for about an hour, and by the time we were finished she and another girl med student had decided to drive to San Diego to see me. They took turns driving Jerry's Chevrolet convertible, and drove straight through in 48 hours. They stayed about a week, at the end of which Jerry and I decided to become engaged. She decided to quit medical school, go to Vincennes and get her belongings, and come to Hawaii in December when we would be married.

After thoroughly testing our twelve new PBY-5 planes, we were ready to fly them back to Hawaii. About noon on 27 October 1941 we began our takeoffs. The weather was great for the first eight hours as we flew in loose echelon, but about an hour after dark we entered the "soup," and we were on instruments until daylight. We flew directly over the first of the destroyers stationed along our flight path, but because we were in the "soup" most of the night we did not see the second and third. The Skipper assigned different altitudes and power settings to each aircraft and suggested that we turn on our landing lights when we estimated we were over the second destroyer. When we turned on our lights, the switch froze and the lights remained on until we landed. This also happened to three or four other planes, and made the night of instrument flying most uncomfortable with the bright lights reflecting in the clouds. The destroyer reported seeing the lights of two airplanes. When we broke out of the instrument conditions about daylight, we could not see another airplane. We were flying at 8,000 feet, and could see the tops of the mountains on the big island of Hawaii. We landed at Kaneohe at 8 a.m., 22.5 hours after we took off. Our planes landed as much as an hour apart, with flight time for the first plane 22 hours, for the last plane 23 hours. We were proud of our new PBY-5s. they could carry heavier loads greater distances, and they were a little faster than the old PBY-1s.

Flying The PBY-5 And 5-A

The average pilot had to have about 1,000 hours before he could qualify as a Patrol Plane Commander in a PBY-5 or a PBY 5-A. Both planes were very versatile,

allowing a high degree of reliability, mobility, and flexibility in performing for the Fleet. Both handled exactly the same in the air, but the PBY 5-A was slightly slower and had slightly less range because of the added weight of the landing gear. However, it had greater flexibility, as it could operate from land or sea. Structurally, the planes were identical with sturdy hulls that could withstand long periods in rough seas and a large high lift wing that would allow takeoffs at weights far in excess of design specifications. The Pratt and Whitney R-1830 engines were among the most reliable ever built, and their location as well as the cowling design allowed exceptionally good engine cooling, so that they would not heat up during long taxi periods on the water or during lean fuel settings in the air.

The PBYs had a maximum endurance of about 30 hours. They were armed with .30 cal machine-guns in the bow, and in a tunnel hatch, and .50 cal machine-guns in each of the two waist hatches. They could carry up to four 1,000-lb bombs, up to four 500-lb depth charges, or two 2,000-lb torpedoes. They were equipped with Norden bombsights that were used in high-altitude horizontal bombing, or which could be used for taking drift sights at low altitude to assist in navigation. They also had a small antenna loop mounted above the hull that could be used as a homing device, or for taking bearings on radio signals, but it was unreliable at distances of more than 50 miles. There were dual instruments in the cockpit for the pilot and the co-pilot.

In discussing the operation of these seaplanes (flying-boats), it is well to recall that pilots got involved in a lot of seamanship, and they had to know a lot about the effect of wind, waves, and currents in the safe and efficient operation of their planes.

Before any pilot entered the cockpit of his seaplane, he should have given it a visual pre-flight inspection of the outside to make sure there were no control locks on ailerons, rudder, or elevators, and that there were no leaks around the engines, or loose cowlings. Once in the cockpit, he made a communication check with all stations on the aircraft to make sure that all radios were working and that each station was manned and ready for operation. He then made sure that the propellers were clear before starting each engine. Once the engines were warm and running smoothly, he gave the beachmaster the signal to launch the plane into the water. Once the beaching gear was removed, the plane was free to taxi to takeoff position. This is where the pilot's seamanship came into play, as he had to use his engines and his flight controls on the water to reach the most advantageous position. By the time he had reached takeoff position, he had made magneto checks of left and right magnetos, checked the freedom of all flight controls, scanned instrument settings, and he and the co-pilot had completed the check-off list for takeoff.

Depending on the wind, waves, and length of the takeoff area, the pilot had to make a judgment that he had reached a safe downwind takeoff position and that he had an ample run ahead in which to get airborne and clear all obstructions. In making this judgment he had to rely on his experience in taking off with various wind, weight, and load conditions. Where loads were heavy, winds were light, and takeoff runs had to be short, pilots developed the technique of making circular takeoffs. Using this procedure, the pilot would begin his takeoff run downwind and get his plane on the step by the time he made a semi-circle and was headed into the wind at an airspeed of 45-50 knots. Thus he did not need much more run to reach the takeoff speed of about 65 knots. Once the plane was airborne, the pilot reduced the power settings for the engines to produce the proper rate of climb and proceeded to carry out his mission. The most critical part of any flight was the takeoff when the engines were at or near maximum power settings; engine failure was most likely to occur during this period. However, engine failure during takeoff was not as critical in a seaplane as in a landplane because the pilot merely cut the throttles and landed the seaplane in the water. In a landplane, he might have obstructions at the end of a runway which would mean a crash. Moreover, the water acted both as a cushion and as a brake on touch down, so there was little roll.

The next most critical time of many flights was the safe landing at the destination, because generally weather could be negotiated without any problems on instruments except for the possible exception of tornadoes, waterspouts, hurricanes, and typoons. The PBYs were the eyes and ears of the Fleet in the years before World War II, as they could scout and patrol for great distances and report the weather or the presence of an enemy. Pilots were well trained for the advanced base operations that were required after the war started. They could operate for long periods of time from aircraft tenders which provided logistic support at great distances from their regular bases. They could also become aggressive if it was necessary to deliver bombs and torpedoes on targets as much as 1,000 miles away, and their depth charges were very effective against enemy submarines.

The PBYs were well equipped for instrument flying, but under such conditions they were ineffective as a scouting or patrol plane if they could not see what was on the surface. These planes were also equipped with automatic pilots (Iron Mikes) which allowed pilots and co-pilots more freedom to observe what was on the surface or in the air ahead of them, as long as the weather was clear. Most patrols were conducted at an altitude of 1,000 feet or less with the assistance of the automatic pilots, but under instrument conditions or low-altitude flight pilots flew manually. One of the most important tasks of a patrol plane was accurate navi-

Opposite page: This shot of the PBY banking past the camera shows clearly its graceful lines and pre-war markings before the red was removed from U.S. roundels and the tail stripes were abandoned.

gation. If a pilot did not know where he was at all times, his flight was not effective and it was impossible to perform his mission. In fact, poor navigation jeopardized the safety of more flights than any other single factor except weather, and when these two circumstances acted in concert they invited disaster. I always considered poor navigation a form of pilot error, because a pilot on patrol who made contact with the enemy and reported his position inaccurately was giving his fleet commander bad information, and thus the admiral either could not act properly on it or acted on misinformation. Furthermore, a pilot in error was not covering the area that his Fleet commander wished patrolled. Lastly, he would have trouble arriving at his destination, if he really was lost. It has been my observation that poor navigation and bad weather have caused the majority of aviation fatalities.

The PBYs were ideally suited to perform the tasks for which they were built because they were slow, reliable, had tremendous endurance, and provided great mobility and flexibility. Fully loaded for combat patrols they would take off at 60-65 knots and cruise at 85-90 knots during the first few hours of flight, but would work up to about 100 knots by the end of a 1,500-mile flight. The planes were well equipped for instrument or night landings, as they had gyro horizons and rate-of-climb and descent indicators. With a rate of descent of 200 feet per minute or less, it was virtually impossible to fly the plane into the water and damage the airplane as long as the wings were level. When the plane reached an altitude of about 10 feet, the pilot could feel the cushioning effect of the wing near the water, and he could chop the throttles and land.

During our operations at Kaneohe after the war started, the entire island of Oahu was blacked out at night and there were no night landing aids in Kaneohe Bay. Fortunately most of the nights were clear, and pilots returning from 1,500-mile patrols could use a small island at the entrance to Kaneohe Bay as a reference point. The island was named Chris Holmes after the owner, who always had his yacht moored into the wind nearby. There was no obstruction light on the mast, so we always made sure that we were well away from the island when we began our landing approach. The prevailing winds were easterly 95 percent of the time, so that we made this landing approach most of the time at night. We would get the altimeter setting from the control tower, reset our altimeters accordingly, and try to begin our landing at 100 feet. A rate of descent of 200 feet per minute allowed us to be on the water in about 30 seconds.

Once we were safely on the water, the flight was over but the mission had not been completed. We had to taxi the plane to the ramp where it was to be received by the beaching crew. The prevailing easterly winds made beaching at Kaneohe relatively easy, even at night, because we approached the ramps heading into the wind and we rarely used sea anchors. Once the beaching crew had the beaching gear firmly attached to the aircraft, a tractor towed the plane "tail first" onto the concrete apron where it was parked and serviced for the next morning's early flight. When we were deployed to the South Pacific and operating from aircraft tenders at advanced bases, we did not have to worry about beaching gear because we did not beach. We tied up to mooring buoys overnight, and the aircraft were serviced there before the next day's early flight. Also, the PBY-5A did not have to worry about beaching gear, as it could simply put the landing gear down and lock it while in the water and taxi up the ramp to a parking place.

The PBYs were about the only planes that could be used for rescue work in the open seas, and special squadrons were designated to train highly skilled pilots in this type of operation. These squadrons were affectionately called "Dumbo" squadrons, and they were responsible for saving the lives of hundreds of pilots and fighting men who were down at sea for one reason or another. These "Dumbo" pilots were the unsung, and mostly undecorated, heroes who operated around the world throughout the war. They developed outstanding skills in landing and in taking off these huge seaplanes in hazardous open sea conditions.

During the period of operations in Jacksonville when we were instructing advanced students in seaplane operations in 1943 and 1944, we conducted all of our water landings, takeoffs, and recovery operations in the St. John's River. There we had the additional factor of current to deal with when we were making approaches to recovery ramps. The river flowed north so that the ideal approach to a ramp was from the north into the current, and if possible, into the wind. However, the prevailing winds there were generally easterly, so we had to use sea anchors while making ramp approaches in order to slow the plane down for the beaching crews. During night operations, one instructor was assigned each night as boat officer with the responsibility of stringing a series of green lights into the wind, and this established the landing course for the instructors who were conducting bounce drill for the benefit of the advanced students. When the wind was directly across the river, it was a little hairy coming in from the west clearing 200-foot trees, landing and taking off and clearing the trees on the other side of the river at the same height where the river was not much more than one mile wide.

For the PBY pilot, a mission to anywhere was not unusual, and most often he accepted the challenge and completed the mission as a routine matter. Although most Patrol Plane Commanders had requested duty as fighter pilots, when they had a choice at Pensacola, most of them became devoted to their beloved CATS (Catalinas built by Consolidated Aircraft Corp.)

A week before we left Coronado, VP-12 pilots had

Opposite page: The PBY cockpit was distinguished by its twin wheels and unique rolling control bar.

flown their old planes to San Diego and were busy testing new planes, and less than a week after we arrived in Kaneohe, VP-14 pilots took off for the mainland to get their new planes. By the end of November 1941 all three squadrons at Kaneohe had new PBY-5s, and each squadron was busy checking out new Patrol Plane Commanders. In our squadron, six pilots — Howe, Coley, Corlett, Poulos, Uebelacker, and myself — all made PPC during November. All three squadrons were involved in Fleet Wing Tactics on 2-4 December with all available planes flying seven to eight hours per day.

While our squadrons at Kaneohe were busy checking out new PPCs and getting ready for Wing Tactics, a secret mission was being planned in Pearl Harbor. Admiral Kimmel and his staff decided that Wake Island should have some airplanes. Civilian construction crews had been preparing a runway there since January, and it was completed in November. To get a squadron of planes to Wake required a lot of planning and directly involved the movement and duties of many officers and men. One of the key officers affected was Cdr. Winfield Scott "Spiv" Cunningham, navigator for the seaplane-tender *Wright*, who had orders to report to Johnston Island as Officer-in-Charge. In mid-November his orders were suddenly changed by dispatch, to Officer-in-Charge of Wake Island, and the *Wright* was ordered to load supplies, materiel, and men to support a Marine Air Squadron. Maj. Paul A. Putnam, commanding officer of Marine Fighter Squadron 211, was ordered to have all squadron personnel and equipment aboard the *Wright* by 19 November, except his 12 pilots and airplanes. Also ordered to report to the *Wright* were Cdr. Campbell Keene, 3 Ensign patrol plane pilots, and 30 enlisted personnel who would support a patrol plane squadron at Wake. One Ensign, Chester Olcott, and 10 enlisted personnel, were assigned from our squadron, VP-11. The other Ensigns and 10 men each were assigned from VP-12 and VP-14. The 47-man ground crew of mechanics from VMF-211 were under the command of Second Lt. Robert J. Conderman. Also on board the *Wright* was Maj. Walter Bayler, a USMC communications specialist who was to set up ground-to-air radio facilities. The *Wright* left Pearl Harbor the morning of 20 November and arrived at Wake Island on Friday morning the 28th.

The mission to deliver the Marine Fighter Squadron to Wake involved a whole Task Force, including the aircraft carrier *Enterprise*, three cruisers, and nine destroyers. Admiral Kimmel selected one of his most aggressive leaders, Vice Adm. William "Bull" Halsey, to command the Task Force and stressed the need for absolute secrecy. When Halsey asked how far he should go if he met Japanese ships, Kimmel advised him to use his common sense. To Halsey this meant putting the carrier under war alert with live warheads on torpedoes and planes fully loaded with bombs and ammunition. Pilots were ordered to shoot down any planes they saw and to sink any ship encountered.

On 27 November Major Putnam was given secret verbal orders to fly his twelve F4F-3 Grumman Wildcats from the Marine Base at Ewa to Ford Island the next morning and then land aboard the *Enterprise*. When one of the fighters failed to start, the pilot was flown to the carrier and assigned one of its F4Fs. On the second day at sea, Admiral Halsey told Putnam they were going to Wake Island. About this same time, a Patrol Squadron at Pearl Harbor was told to fly to Midway, then arrive at Wake on 3 December to fly six-plane search patrols. The scouting search patterns flown by the squadron were designed to protect the Task Force. The twelve planes fanned out to cover a wide area between Midway and Wake on 3 December. Then on the 4th, five planes searched areas to the west of Wake, while one flew northeast to meet the Task Force and escort the fighters to the landing field. The rendezvous was made about 200 miles from the island, and the carrier began launching the fighters. The PBY led them to the field where they zoomed the 1,700 people (all hands) lining the runway before coming in for perfect landings. It was a cheering sight for the civilians and military alike and gave them a feeling of protection when they saw the four .50 cal machine-guns extending from the wings. However, the problems for VMF-211 had just begun. Few of the pilots in the squadron had more than thirty hours time in the plane, and the aviation section under Lieutenant Conderman had inexperienced mechanics unfamiliar with F4F instruction manuals left behind at Ewa. Furthermore, the 100-lb bombs would not fit the bomb racks, but through the efforts of ordnance officer Capt. Herbert C. Freuler and 1st Lt. John F. Kinney, they fashioned support bands out of sheet metal, allowing them to hang two 100-lb bombs on each plane. Kinney was a Marine aviation cadet in 117 C. Putnam established combat air patrols of four planes each, circling the island at a distance of fifty miles. These patrols were flown at dawn and dusk each day, and the first patrol took off at dawn on 5 December. With the combat air patrols established, the PBYs made their last six-plane patrol to the west that same day, and then all twelve planes took off for Midway the next day, arriving the same date as they crossed the International Date Line. They landed back at Pearl Harbor the afternoon of 6 December.

After the *Enterprise* launched the twelve fighter planes, Halsey turned his Task Force around and headed back to Pearl Harbor. At a Force speed of 20 knots, he expected to have his squadrons there by 8 a.m., December 7th, and have his ships in harbor by noon. He was well pleased with the manner in which he had performed his mission. Although he was ready for any eventuality, his pilots did not see a ship or an airplane, except the friendly PBY that came out to meet them. He had delivered the Marine fighters secretly and safely to the point where they were launched, and he had carried out his orders from Admiral Kimmel successfully.

Although the delivery of twelve fighter planes and pilots to an island 2,100 miles away does not sound like much of an operation today, it was quite an undertaking in 1941. It involved the lives and duties of thousands of people in the Navy of a country which was on the brink of war, and many of these people were Reserve pilots who were trained as Aviation Cadets. Probably more than 50 percent of pilots on the *Enterprise,* in the PBY squadron, and in the Marine Fighter squadron were Cadets.

I was "sitting on top of the world." I was a newly qualified Patrol Plane Commander in a squadron with brand new airplanes. I was also engaged to be married, and my wife to be was scheduled to arrive in Hawaii on 11 December. I had rented a small place to live in Kailua. On the 5th I received a call from Fred Roehm, a Delta Upsilon fraternity brother at IU, that his supply ship had just pulled into Pearl Harbor for the weekend. Fred had wanted to be a pilot, but knew that he could not pass the eye test. His eyes were good enough to pass the test for the Reserve Officer training program, however, and he became a Reserve Ensign early in 1941. I had not seen Fred for 2½ years, when we graduated at IU on 6 June 1939. He was free for the weekend, and I was to meet him at the Officer's Club in the Pearl Harbor Navy Yard for lunch on Saturday. I then we would spend the afternoon at the Outrigger Club at Waikiki and the evening in Honolulu and on the beach.

Fred and I met at noon at the Officer's Club, had lunch and a few drinks, and then went to a friend's apartment where we changed into swimming trunks for the beach. We saw several of my friends at the Outrigger Club, and they agreed to meet us for dinner at 7 p.m. at Lau Ye Chau's restaurant (we called it Lousy Chow's). We had a great time at the beach and a good dinner at Chau's, then spent the rest of the evening in and out of bars around Waikiki. When the bars closed, we were underneath the banyan tree outside the Moana Hotel. Although we had been drinking for more than twelve hours, and hardly needed another, we headed for Society Sal's. She was the Madam in town who controlled all of the girls, and even if we had no interest in girls, Sal was happy to serve us a good drink after hours at $1 each. We had two drinks each with Sal, and then we were on our way to the Pali and the treacherous road that led us to the BOQ at Kaneohe. We were in the "sack" by 3 a.m. 7 December 1941.

Chapter Four

Where They Were On Pearl Harbor Day

Although Fred Roehm and I had started 7 December 1941 under the banyan tree at the Moana hotel, and had assured Society Sal about 2 a.m. that a Japanese attack on Hawaii was a million to one shot, we were rudely awakened on the top deck (third floor) of the Kaneohe BOQ a few minutes before 8 o'clock. There was an unusual amount of racket for this time of day on Sunday, so I got up and peeped out one slat of the venetian blind, which looked out over the bay and the hangar area. I saw several planes buzzing around and some smoke coming up from the apron around the hangars, but I told Fred to go back to bed, it was the Army "fly-boys" trying to wake us up. I went back to bed, but there was much excited shouting in the passageways, so I got up again and raised the entire blind. What I saw was a plane strafing the BOQ and the red "meat balls" as it pulled up to clear the roof. I said, "Fred, this it it! The Japs are here!" We grabbed our clothes and ran downstairs to the first deck. Somehow we thought this would be a safer place to dress, because we expected the bombs to be dropping at any moment.

We were dressed in nothing flat, and made it to my car parked outside the BOQ. Japanese planes were strafing us all the way to the main gate. I told Fred he would have to make it to his ship from there. (I later found that he commandeered a milk truck at the gate which took him to his ship in Pearl Harbor). I drove to the hangar area, as fighter planes were strafing me all of the way. I parked about two blocks from the hangar and ran down to the road adjacent to it where I saw our Skipper, most of our officers who lived in the BOQ, and many of our enlisted personnel who lived on the Air Station. By this time it was about 0830. The Japanese dive bombers and fighters had expended their bombs and ammunition and were joining up to head back to their carriers.

All three Commanding Officers of VP-11, VP-12, and VP-14 were mustering officers and enlisted men and giving them instructions on the road adjacent to their respective hangars. Captain "Slim" Johnson was very calm and collected and advised us in VP-11 to first look after the wounded and the dead. He said there were many people missing, there were many casualties, and there would be many more, as the next wave of planes would be horizontal bombers going for the hangars. All officers and chief petty officers were to report to the armory on the double and draw .45's with ammunition belts, and Springfield rifles, then report back to the same area. All other personnel were to go to the hangar and salvage as many records as possible, bringing them back to the area in any type of containers they could find. He stressed the fact that all hands should get as far from the hangars as possible when the bombers were sighted, and this warning saved many lives during the next attack. There was about 40 minutes between the end of the first attack and the arrival of the second. We estimated there were 12 dive-bombers and 12 fighter planes in the first attack, and they did an incredible amount of damage to our aircraft and hangars. Most of the airplanes parked on the concrete aprons outside the hangars were either on fire or filled with holes by the fragmentation bombs and strafing attacks.

Ensign Rodney Foss, the Duty Officer for 6 December, was due to be relieved in eight minutes on 7 December by Ensign Joe Smartt, when the Zeroes began their strafing attacks. Ensign Foss was struck by a 20 mm cannon shell on the first run at 0752 and was killed instantly, to become the first casualty that day. Ensign Smartt was in the hangar getting ready to relieve Foss, and when he reached the Duty Officer's Station he saw that Foss was dead. He assumed his duties as Officer of the Day immediately, but he was killed in just little more than an hour when the medium-level bombers dropped a cluster of bombs directly on the VP-11 hangar. As incredible as it sounds, our squadron lost two duty officers in the first hour of war. The toll for our squadron that day was 7 dead and 20 wounded. The total for the air station was 19 dead and 66 wounded.

During the first muster outside the hangar at the end of the first attack, there was much confusion, shock, and anger, as we began counting noses. I could account for many of the missing pilots. I believe Willie Howe, who had the third deck room next to mine (we shared the bath in between), was spending the weekend with his mother and two sisters at an apartment they had rented near Diamond Head. I did not expect to see him at the first muster, nor did I expect to see any of the married pilots who lived outside of the Air Station. I did expect to see Charles Willis, because he, George Clute, and Hubert Reese left the BOQ in his open touring car about the time Fred and I left for the main gate. On the way to the hangar, Willis stopped the car, and Clute and Reese hit the road running. This was the last anyone

Taken from Aeia Heights, one of the first photographs of the attack on Pearl Harbor shows the ***Arizona*** *on fire on battleship row, Ford Island. (USAF)*

saw of Willis, and we did not find out what happened until the next day. After Clute and Reese left the car, Willis continued toward the hangars after restarting his car. A bomb from a dive-bomber hit the car and blew him into a thicket along the side of the road. The car continued down the road without Willis, hit a utility pole, and burst into flame. Among the burned remains of the wreckage were many charred human bones, enough for several people. We all thought Willis and possibly others had burned to death, but we learned the next day that he and Reese had discovered a Hawaiian burial ground near the beach and had collected some bones which they had put in the back seat on Saturday. Because no one had seen Willis, we assumed that he was dead and that his were among the bones discovered in the wreckage. Although there was no positive identification, the Navy notified his parents that he was "missing in action." When Willis regained consciousness in sick bay, on 9 December, he was allowed to telephone his parents that he was okay and no longer missing.

I was standing at the north end of our hangar shortly after 9 a.m., when I spotted the bombers coming over the mountains. I shouted the alarm for all personnel to clear the hangar, and began running as fast as I could away from it. I hadn't made 100 yards before the fighters began strafing to clear the way for the bombers. In this area I found some large spools of steel cable and some heavy construction equipment and decided this was a fine place to wait out the attack. The bombers must have been at an altitude of 3,000 to 5,000 feet, and there were about eight three-plane sections. I could easily spot their clusters of bombs as they were dropped, and most of them landed directly on the hangars, but a few landed wide of their targets in the bay as their runs were made in a strong cross-wind. When the bombs began exploding, the spools of cable started dancing around, and I thought I had made a bad choice of spots, but there was no problem. Once all bombs had been dropped, the bombers joined the fighters in strafing everything in sight. We picked up our rifles and began firing back. Before the planes expended all of their ammunition, we had shot down one fighter plane whose pilot made his last run. He was DOA (Dead On Arrival) as he made no attempt to pull up. His plane hit the ground about two thirds of the way up the hill toward the BOQ, and his engine rolled to a stop about 50 yards from the back door. Our machine-gun and rifle fire had punctured the external gas tanks of two bombers which were streaming gasoline vapors as they headed back for their carriers, and we guessed they would not make it. The last of the attackers was out of sight by 9:30 a.m., and we were all in a complete state of shock. There was devastation in all directions. None of our planes were operational, the VP-12 and VP-14 hangars were demolished, and more than one-half of our hangars could not be used. The dispensary was overflowing with dead and wounded personnel, and emergency care rooms had to be set up in the administration building.

Most of the officers and men who lived off base had reported for duty by the time the Skipper was assembling all hands at 1000. He told us we probably had not seen the last attack and that there was a good possibility of an invasion of the islands. He said, "We don't have any airplanes that will fly, but we are professionals at war, being paid to fight to the best of our ability with what we have." All ordnance men were to concentrate their efforts in getting machine-guns out of wrecked planes and setting up emplacements at strategic points along the beach. All personnel with rifles were to go to the armory and get five clips of ammunition (100 rounds). When he learned that Joe Smartt had been killed, he appointed Jack Coley as the new Duty Officer and ordered him to set up four-hour watches around the hangars, at all machine-gun emplacements, and at all telephone stations in the barracks and the BOQ. Most of the telephones on the base were working, except those that had been demolished in the hangars. The Army had responsibility for all communications in emergencies such as this, and they assumed control of all radio broadcasts. There had been radio reports that Japanese parachutists had landed, and this suggested that an invasion had begun. Emergency rations of cold sandwiches and hot coffee were being served around the clock in the general mess and at the BOQ.

Because invasion appeared to be imminent, the Commanding Officer of the air station, Cdr. Harold "Beauty" Martin, ordered every piece of equipment that would move to be placed strategically on the air strip, including private automobiles, construction equipment, and all military equipment that was not in use, in order to make it impossible for enemy planes to land. Commander Martin also decided that sailors in white uniforms would present invasion forces with easy targets, so he ordered them to dip their whites in large vats of coffee which were set up in several places around the station. This made the uniforms look like dirty tan fatigues, and made the sailors most uncomfortable, but it also made them much less conspicuous.

The station Gunnery Officer, LCdr. William "Red" Raborn wanted to clear the armory of all small weapons, so he issued the last of the Springfield rifles and Colt .45 pistols, along with the five clips of ammunition, to ordnance men throughout the station. Once I replenished my ammunition at the armory, I went to my car and gave it a thorough examination. Although I had been strafed by planes on the way to the main gate and the hangar, I could only find three bullet holes, and there was no damage to any of the moving parts.

While driving to the landing field to park the car, I began to wonder about Jerry Ringham, the girl I was engaged to marry. The last time we had talked was 3 December, and she was leaving Vincennes the next day, arriving in Los Angeles Sunday morning by train. Her ship was to leave Monday, 8 December, and she was to arrive in Honolulu Saturday morning, 13 December. I found out later that the newsboys were yelling, "EXTRA! Japs Bomb Pearl Habor!," as she stepped off the train in Los Angeles. She remained on the West Coast for a week, trying to find a way to get to Hawaii, but Naval authorities assured her it was not possible. They told her, in fact, that they were planning to evacuate all dependents from the islands, so she got a train ticket back to Vincennes. Although I wrote to her several times during the next two weeks, it was several weeks before we talked by phone.

Earlier while Fred and I were sleeping, three PBYs from VP-14 were taking off at first light for their dawn patrols. They were anti-submarine search patrols, armed with four 250-lb depth charges. Each plane had an assigned search area which extended from a starting point on Oahu. The eastern sector started at Makapuu Point and covered the western shores of the islands of Maui and Lanai, Lahaina Roads, and areas 100 miles south. The middle sector covered an area southwest of the Pearl Harbor entrance channel out to 200 miles, and the third sector covered an area to the northwest about 50 miles beyond Niihau. On this Sunday all three sectors were covered by Ensigns who were Reserve pilots trained as aviation cadets. Classmate Jack Roberts, 130 C, had the eastern sector; William "Tiger" Tanner, 129 C, had the middle sector; and I believe Freddie Meyer, 107 C, had the northwest sector. At about 0630, Tanner and his crew spotted an unidentified submarine off the entrance channel to Pearl Harbor. Ten minutes later, they had dropped their four depth charges and observed the destroyer *Ward* delivering depth bombs. Tanner sent his encoded report to CINPAC and Kaneohe as follows: "Attacked and sank unidentified sub operating in defensive sea area." He then continued on his patrol, hoping and praying he had not sunk a friendly sub. All three planes continued on assigned patrols until they were diverted to search areas northward after the first Japanese attack. Fuel limitations prevented them from making contact with the main enemy force, but one plane crossed paths with Japanese returning to their carriers who had put enough holes in his plane to force him to beach it at Kaneohe. When Tanner and Roberts returned that afternoon, they had the only flyable planes out of 36 new PBYs at the Air Station.

Although Admiral Halsey's Task Force had expected to be in Pearl Harbor by 0800, rough weather delayed destroyer refueling until daylight and the Force was four hours behind schedule. Halsey launched an early morning scouting flight at 0600 to sweep a 180 degree sector ahead of the force, with instructions to land at Ford Island. Thirteen planes from Scouting 6,

B-17s at Hickam Field after the attack with burning hangars in the background. (USAF)

and five from Bombing 6, were ending their sweep as they approached Oahu between 0745 and 0815. Unfortunately, the first Japanese fighters began making their runs at about the same time and began shooting down *Enterprise* planes flying between Barber's Point and the Marine base at Ewa. Two of my good friends who were regulars in the BOQ poker games joined in formation and were approaching Ewa when they were hit. "Mannie" Gonzales was hit on the first run, and his buddy, Norman "Westie" West, said, "I saw him slump over in the cockpit. He never had a chance to bail out." In addition to Mannie, Ensigns John Vogt and Walter Willis were killed when they had no chance to bail out. Lt. (jg) Clarence Dickinson and Ensign Bud McCarthy had their planes shot out from under them, but parachuted to safety. Ensign Ed Deacon's plane was shot down by U.S. anti-aircraft fire, but he managed to land in the ocean near the shore, and he and the rear-seat man survived. All other pilots landed safely at Ewa or Ford Island, except Lieutenant (jg) Patriarca who crash-landed on Kauai after he ran out of fuel. Most of them, however, sustained some damage from the Japanese or our own A/A fire. For the rest of the pilots in the Air Group, it would be a long time before they set foot on Ford Island. The fighters were sent up on continuous combat air patrol, while the scouts, bombers, and torpedo planes were sent on extended searches. One torpedo pilot, Ensign "Bucky" Walters, overextended his search, could not find the *Enterprise*, and decided to try for Kaneohe. He wisely avoided Ford Island, as he correctly suspected that all hands would be firing at everything that moved. It was about an hour after dark when he approached the blacked-out field, but he was smart enough to have his running lights on (red on the left and green on the right). Everyone on the Air Station knew it was an American torpedo plane, but not one person would guarantee that the pilot was American. He made one pass at the field with his landing lights on, and I expect he wished he had gone to Ford Island. The amazing part about this whole episode was that not a shot was fired. Almost everyone had a hand in preparing the field for the night, and we were sure that no plane could land. But on the next pass, Bucky made a perfect landing with his 2,000-lb. live torpedo.

Because there had been no attacks made after the first two, and no invasion attempt had been made, it was decided at noon that a situation report should be given to Admiral Bellinger. Our Skipper, LCdr. "Slim" Johnson was selected to make the report, and he spent much of the afternoon at Ford Island. He returned to Kaneohe about 1700 and held his last muster of our squadron. He told us he had received verbal orders to Staff duty on Ford Island, and that he would be leaving the next morning. He also told us that all patrols would be run from there, and that most of the pilots in all three squadrons would be flown there in the flyable planes early in the morning. He said that Lt. F. R. "Snooks" Jones would be in command of the squadron until further notice and that we were to give him our very best efforts as professional Navy men. He said that an invasion was quite likely and that all hands were to observe a strict "black-out" after dark (not even a cigarette could be lighted or smoked). He advised all hands assigned watches to get near their posts, and all others to get where they wanted to be before dark, because they should not be moving around in total darkness. He pointed out that everyone with a firearm was "trigger-happy," and if we began moving around in the dark, we would be firing at each other instead of the enemy. He said he was sorry he had to leave such an outstanding group, gave us a farewell salute, and wished us good luck. When we were dismissed, I asked, "Captain, how bad is it at Pearl?" He said, "Joe, it's so bad I can't talk about it. It even makes me ill to think about it!" There were tears in his eyes when he said it.

After the last squadron muster, Hubert Reese, Bill Howe, and I went back to look over the charred remains of Willis' car. Reese told how he and Clute had started out with Willis from the BOQ, but they jumped out when he stopped the car and began running in different directions. That was the last they saw of him. We then proceeded to the BOQ to decide what we were going to do, and where we were going to spend the night. None of us had any watches, but I was scheduled to be the Duty Officer the next day and would relieve Jack Coley at 0800. Most of the married officers who lived off base were invited to spend the night in married officer's quarters on the base or in spare beds in the BOQ.

About an hour before dark, Reese, Howe, and I decided to make our battle stations about 100 feet from the back door of the BOQ, on the hillside overlooking the bay and the beaches surrounding the hangar areas. We thought this would give us the greatest amount of warning time in case of invasion, and we all three vowed we would fight until our last breath. We would not be taken prisoner. This station was also close to hot coffee and sandwiches, and we informed the guard on the door that we would be coming in and out during the night. We got our rain gear (it was misting at the time), got a cup of hot coffee and a sandwich from the galley, and established our vigil on the hillside. We had not eaten anything all day, and we were getting hungry. Bill Howe (bless him) brought a bottle of gin, and we had laced coffee until midnight. The three of us talked all night, speculating what we would do if the Japanese landed on the beaches, and what would probably happen to us even if they did not invade us. We talked about our families. Bill Howe, I think, had just left his mother, and sisters, Anne and Barbara, that morning in the rented house at Diamond Head. They did not know if he was still alive or if they would see him again. Hubert, the only child of parents who lived in Memphis, wondered how long it would be before they knew he was alive, if he survived. My parents lived in Bruceville, Indiana, and the last time I had seen them was when they came to Jacksonville when we delivered our old planes. Also, I was wondering what happened to Jerry Ringham in Los Angeles that morning, and if we would ever get married, or even see each other again. Bill Howe and I wished we had gone to China with the AVGs. We discussed the odds of making it through the war, which we concluded would be a long one. If we were not killed in the next few days, we figured the squadron would be deployed toward the west and Japan, and the lumbering PBYs did not have a chance in aerial combat. We assessed the odds at about ten to one that we would not make it. As it turned out, we were a little harsh in our assessment, though I was the only one that made it. We took turns going to the galley for hot coffee, and this gave the runner a chance to smoke a cigarette in the passageway by the galley. Even so, it was a very long night!

I have covered most of the things I can remember about NAS Kaneohe on this day of infamy. Just about every person who was old enough to remember anything at the time remembers where they were and what they were doing when they first learned about Pearl Harbor. Several of the senior Reserve Ensigns in all three Kaneohe squadrons were enjoying leave with their families back in the States, including Kauffman, Heyer, Pennell, and Bankhardt from VP-11. They had completed more than the required two years in the Fleet and had received orders in mid-November to report to E Base as flight instructors. I'm sure they were concerned about their former shipmates, but they were quite happy to be where they were.

At Wake Island it was 0650 on 8 December, when Major Devereux's radio operator tuned into the Hickam Field frequency and heard in the clear (plain English) that Hickam was being attacked by Japanese dive-bombers, and this is no drill! The bugler sounded general quarters, and all hands went to their battle stations. Major Putnam got the word just before his two pilots took off on morning patrol, and he sent up four planes instead of two, as he led the flight. Four planes were standing by for takeoff, and four were undergoing maintenance checks. Commander Cunningham de-

One of the midget Japanese submarines beached near Bellows Field, Hawaii, similar to that bombed and attacked off the entrance to Pearl Harbor shortly before the carrier strikes began. (USAF)

cided he should recall Captain Hamilton and his Boeing Clipper which had taken off an hour before, headed for Guam. By 0930 the flying-boat was back in the lagoon. The four fighters on patrol had returned, and the four relief planes had taken off.

Gregory J. W. Urwin, in his *The Wildcats of Wake Island,* indicates that the second combat air patrol was making a sweep north of Wake at 12,000 feet when 36 Mitsubishi medium bombers (Bettys) began their attack with 100-lb bombs and 20 mm incendiary shells. Four Marine pilots tried to get to their planes and get them in the air, but three of them, Holden, Graves, and Conderman died in the effort, while Spider Webb was seriously injured. In just a few minutes, VMF-211 had lost seven F4Fs, and the auxiliary gas tank of the other plane on the ground had been ruptured. Twenty men had been killed, including the three pilots above, and eleven had been wounded, including Webb and three other pilots. When the CAP in the air returned to land, the pilots could not believe the damage that had been inflicted. Three of the planes landed safely, but Henry "Baron" Elrod damaged the propeller and engine mount on his plane when he landed among the bomb debris. The air defense was cut to three planes, and two that could be repaired. The new engineering officer, John "Mother Necessity" Kinney, and his chief assistant, Technical Sergeant Hamilton, worked all afternoon and night getting the two cripples ready to fly.

Although the Boeing Clipper in the lagoon had many bullet holes in it, Captain Hamilton and crew decided that it would fly. They also decided that Pan Am would cease operations on Wake and fly the 29-man detachment and passengers to Midway. The plane was stripped of everything but people and aviation gas, but the Clipper pilot had to make three tries before he got the plane in the air.

The pilots and planes of VP-22 had just completed their longest advanced base operation in history. They had spent seven weeks at Midway, operating from an aircraft-tender. They flew search sectors every day until 3 December when they flew advanced patrols for Halsey's Task Force which was going to Wake. They flew search patrols out of Wake on the 4th and 5th, with one plane escorting the VMF-211 fighters to the island on the 4th. They left Wake on the 5th and arrived back at Ford Island still on the 5th and thus had one day's rest before the attack. Because of their extensive advanced base operations, VP-22 was one of the first squadrons to be deployed southwestward.

Ensign Harvey Hop was happy to be back with his wife Miriam, and they spent the afternoon of the 6th shopping in Honolulu. Harvey bought a new sport coat for $35 (quite an expenditure for him at the time), but he never got to wear it, as the war started the next day and he was in uniform until it was over. The coat was given to Russian relief a year later with $100 Harvey had secreted in one of its pockets. He often wondered how the lucky Russian spent it.

Ensign Norman Brady, a classmate in 130 C and a new PPC in VP-23, had his first date with a girl named "Noni" the night of the 6th and did not get back to the Ford Island BOQ until early morning. He might have gone to bed earlier, if he had known what he had to face at 0800 as the Tower Duty Officer. But in looking back, he says he would have done the same thing, because "Noni" eventually became his wife. Also in VP-23, Orville Kellett in plane 23-P-6 flew a search mission for his squadron that day. He encountered two Zeroes on floats, but they did not bother him and he did not bother them. His plane was the only one in the squadron that was not destroyed.

Don "Pop" Korn, a Delta Upsilon fraternity brother of mine at Indiana University (128 C), was Officer of the Deck on the Cruiser *Raleigh* (CL-7). He had just relieved the previous deck officer, when he observed a three-plane formation drop three torpedoes, one on the *Utah,* one on the *Detroit,* and one for his ship. He thought the Army fliers were really going "bananas" until the torpedoes exploded. After the attack the *Raleigh* was upright, but resting in the mud. Don later turned the "deck" over to another officer and got his SOC in the air, but his search pattern was 180 degrees away from the enemy. He did come back with several holes in his tail, obviously from friendly guns. After temporary repairs at Pearl, the *Raleigh* had to go back to Mare Island for extensive repairs and overhaul.

Ensign Jim "Andy" Hardy got a different view of the attack. He had spent the night in town and was standing on Landing C waiting for a boat to take him to Ford Island. He saw the planes push over in their dives, and thought they were from the *Enterprise* until their bombs exploded on the hangars. Then he saw the red "meatballs" on the torpedo planes which were flying up the channel with flaps and landing gear extended to slow them down so their "fish" could not miss. The boat did not come to the landing until the attacks were finished.

Ensign Mark Wood (129 C) was the SOP (Senior Officer Present) in VJ-3 Utility unit operating from Maui. The unit was super-secret, but was ordered to search a sector 200 miles out from Pearl and return. The unit was a radio control group consisting of one SNB (Beechcraft), two SBD (scout bombers), and two J2F (amphibians). Mark led the flight in the SNB while the other four planes were flown by enlisted pilots. All planes were without armament. They saw nothing except six of our destroyers which opened up with A/A fire. None of the planes were hit until they returned to Maui and were fired on by a National Guard unit established to protect the field. All of the pilots were recommended for the Navy Cross, but Mark said, "As a group, we declined it, because we did not feel that we deserved it."

Many of the Reserve Ensign pilots were in the Ford Island BOQ when the attack started, including Paul Hart (128 C) and "Cloudy" Riley (131 C) who were two of the four officers that came to Honolulu with me on the *Matsonia* in August 1940. About 40 officers were in the BOQ, and most were strafed as they tried to make it to their battle stations in the hangar areas.

Ensign "Andy" Reid (128 C) was a pilot in VP-102 which was assigned to Patrol Wing 10 in the Philippines. The squadron had planes anchored in Subic Bay, at Cube Point, and Sangley Point, where their pilots had just managed to bring the planes safely through a severe typhoon. They put "spoilers" (control locks) on all surface controls (rudders, elevators, and ailerons) and filled the hulls with water, up to the floor of the cockpit. Then they fastened the planes to mooring buoys by cable and taxied them at the buoys until the typhoon passed. But this was much worse than a typhoon. The Operations officer awakened Reid and told him that the Japanese had bombed Pearl Harbor. Ensign Jack Grayson, a classmate in 130 C, was also in the Philippines in VP-102, but he was at Ologapo. Both of these fellows, as well as the rest of the pilots in the squadron, knew it would be a matter of hours, or perhaps days, until they would be overrun by the Japanese. Andy eventually became a POW and had to endure the "Death March," but Jack and his crew retreated to the Dutch East Indies where he was shot down the first time while on a bombing raid on the Celebes island of Menado. Jack and his crew spent seven days drifting in life rafts until they were picked up by classmate Jack Davis (130 C) who had come south from VP-24 at Ford Island.

On the other side of the world, VP-72 was stationed at Argentia, Newfoundland. The pilots were safe from Japanese attack, but they were flying anti-sub patrols for convoys in the North Atlantic under severe and adverse weather conditions. One of their pilots, Ensign Jack Poteat (150 C), had to make a completely blind let down in the bay on 6 December. When the squadron got the word that the war had started, they also got orders that the entire squadron would fly to San Diego, and then to Kaneohe to fly war patrols for Patrol Wing One.

Ensign Frank "DeLo" DeLorenzo was in his "snake" ranch in Coronado with six other Ensigns and their girl friends. They were getting "revved up" with Bloody Marys before tuning in a pro football game, when they heard on the radio that the Japanese had bombed Pearl. All of the pilots reported to their hangar, and DeLo was assigned a 600-mile patrol south along the coast in his four-engine Coronado flying-boat.

Shortly after takeoff he heard another plane on his frequency say he had a sub trapped a few miles south of Coronado. DeLo circled the area, but there was no sign of a sub, or a plane, and he did not see anything unusual on the 1,200-mile flight.

DeLo's wasn't the only "snake" ranch where Reserve Ensign pilots were getting "revved up" when they heard on the radio that the war had started. Doc Burns (97 C) and three others were at his ranch with girl friends, and it was just outside of the main gate at Corpus Christi. Bill Chester (135 C), Bill Harrington (133 C), and Bill Graham (136 C), all had ranches in Jacksonville, outside the Naval Air Station where they worked as flight instructors. For all of them, this would be their last Sunday off for the duration of the war. The same was true for the flight instructors at Pensacola.

For the Flying Tigers, this day of infamy was no worse than any other day. They were still in China fighting the Japanese and getting paid for it. In fact, things were looking brighter for them, because they were told they would get $500 for each confirmed Japanese plane destroyed on the ground.

The official status of the American pilots flying with the British Coastal Command was changed from "Observers" to U.S. pilots. They could not receive decorations for their bravery, and they would no longer be shot as spies, if captured.

Navy and Marine pilots who were ex-cadets were involved in every phase of Naval and Marine Corps operations around the world. About 70 percent of the pilots in the air groups assigned to our carriers were cadet trained and were aboard the *Yorktown* in the Atlantic Fleet, the *Lexington* and *Enterprise* in the Pacific Fleet, the *Ranger* and *Saratoga* on the West Coast, or involved in "shakedown" cruises on the *Hornet* and *Wasp* which would be sent to the Pacific. The *Yorktown* was sent to the Pacific, and the *Saratoga* would be leaving the West Coast for Hawaii in two days. The *Langley* was assigned to the Asiatic Fleet, but its status had been reduced to that of a seaplane-tender rather than a carrier. About 75 percent of our patrol-plane pilots were ex-cadets assigned to squadrons in the Philippines, Alaska, Pearl Harbor, Panama, Newfoundland, and the East and West Coasts, and about the same percentage of cruiser and battleship pilots were cadet trained.

Most of the early cadets who had completed their four-year contracts got out of the service and went with the airlines or became test pilots for aircraft manufacturers. For the most part, they were allowed to remain with their civilian employers, but some of them were called back to service with transport squadrons. Others who were instructors at Naval Air Reserve Bases and Air Stations were allowed to remain on active duty as Reserve officers.

Very few Reserve pilots were on active duty in the Washington, D.C., area when the war started, because they lacked the seniority required for duty in the Bureaus. Charlie Harbert (107 C) was the Duty Officer at NAS Anacostia when he got the word about the Japanese. He called the Commanding Officer and the Executive Officer, then doubled all watches, and started breaking out the guns and ammunition. No one could be sure of what would happen next.

I did not see Society Sal again after the war started, but I'm sure that Martial Law and the evening curfews in Hawaii changed her operation completely so that her charges were sleeping at night and working in the daytime.

Chapter Five
Wartime Operations

The long night of 7 December and early morning of 8 December had passed without a single shot being fired at Kaneohe. Although I had less than five hours sleep the night before, and none during the past 24 hours, I did not seem to be tired. I went to the BOQ galley, had a sandwich and a cup of coffee, and headed for the hangar to relieve Jack Coley as the Duty Officer. In the galley I saw Ensign "Buckey" Walters and learned that he was the torpedo pilot that came in to land without a scratch on a strip filled with automobiles and bulldozers. Buckey was an E Base classmate at St. Louis, number one in the class, and went to Pensacola a class ahead of the rest of us, in 129 C. He said he was interested in getting his plane back to Ford Island as soon as possible. I told him we were sending some PBYs to Pearl about 1000, and we would have the pilots notify the tower that a Torpedo plane would be in later. I also told him I would get some men to move equipment from the runway so that he could get his plane off.

After relieving Jack, the first order of the day was to check to see that next of kin had been notified about the dead and wounded. By 1000 we had repaired four PBYs in the three squadrons, and with the two good planes brought in by VP-14's Tanner and Roberts, Kaneohe had six flyable planes to send to Pearl. Most of the pilots in the three squadrons, except the top echelon, the engineering officers, and the material officers went to Pearl in the six planes to fly war patrols from there. By 1100 we had one side of the field cleared so that Buckey could take off. After he did so, he made a low pass at the field, waving to all hands, and that was the last time I saw him. He was killed in the Battle of Midway. The rest of the day was spent salvaging squadron records, engine and plane spare parts, and working to repair airplanes. Not a shot had been fired on the Air Station, as the second wave of attacking planes began to return to their carriers, minus one fighter plane and probably two damaged bombers. However, this situation was about to change, when about two hours after dark, someone on the beach fired a red flare, and all hell broke loose. Everyone that had a gun began firing at imaginary invasion troops. The Wing Commander got me on the phone and asked, "What the hell is going on, Hill?" I told him it looked like the invasion had started, but I would check it out and call him back. I then hit the deck, as bullets and tracers were flying in all directions. As suddenly as the firing started, it stopped after about ten minutes, and I called the Wing Commander and told him there was no invasion and that the firing was triggered by a red flare along the beach. I told him I would check for casualties and report back. After checking for more than an hour, I could find no dead or wounded, and so reported. The rest of the night remained calm, and there was no invasion.

The planes we sent to Ford Island were immediately put to work. They, as well as some flyable planes at the Air Station, were loaded with 500-lb bombs and were sent out on night high-altitude search and reconnaissance missions with orders to drop bombs in case of contact. The planes were launched just before dark in three-plane sections, with orders to search sectors to the west and northwest. At least two of the sections reported contacts and dropped their bombs. Some of our surface forces reported that they had been attacked by high-altitude bombers and that bombs had straddled one force with two near misses, but no structural damage was apparent. These reports of contacts and attacks strengthened the beliefs of our leaders that invasion forces were not far away.

Two of my good friends were lost on one of these night missions, just before they made their bomb drops. Paul Hart (129 C) and Tex Stallcup (133 C) were flying right wing in a section at 10,000 feet when their plane went out of control. The left section pilots said the last time they saw the plane it was in a vertical dive; nothing was ever recovered.

Before the first week ended, Wing and Fleet commanders concluded that there were no invasion forces and that our bombers and surface forces had been firing at, and attacking, each other. As more planes became flyable, they were sent on long-range day patrols out 700 miles, across 100 miles at end of sector, and back 700 miles. Also during the first week of operations, three planes were sent to operate from the aircraft-tender *McFarland* (a converted World War I four-stacker destroyer), anchored in the bay at Hilo, on the big island of Hawaii. These planes ran daily search patrols around the islands to make sure there was no evidence that Japanese submarines had landed invasion troops. The PBYs were not the only bombers sent on high-altitude search and attack missions, as several Army B-17s from Hickman Field were sent on such night missions during the first week. Although several hundred bombs were dropped on our surface

*DD-336 **Litchfield,** a sister ship of the **McFarland**, steaming off Panama before the war in the good old days when smoke was impressive. (USAF/**Aerospace Historian**)*

vessels at night, not one of our ships was damaged, and none of the bombers were damaged by anti-aircraft fire except possibly Paul Hart's plane.

By the beginning of the second week of the war, things began looking up at Ford Island, when the carrier *Saratoga* brought some new fighter planes, and a fresh new air group to the Pacific. This increased our carrier strength by 50 percent, and gave all military forces in the islands a big boost in morale. Ensign Harvey Hop of VP-22 assumed the Tower Duty at 0800 on 14 December, and later that day he was able to get his wife Miriam to visit the Ford Island Tower. (Pilots could not go ashore but dependents could visit until 1600, when the word was passed for all dependents to leave.) Harvey and Miriam were busy that afternoon recording on paper what happened on 7 December, and did not hear the orders, "All dependents ashore." By the time they got everything on paper, it was too late for Miriam to leave and get home before dark. (Any person apprehended after sundown was placed under military arrest.) They didn't want Miriam to get arrested, so they decided she would have to RON (remain over night). Harvey got her a sleeping bag and she slept under the control console counter, undisturbed by the Admiral's tower inspection that night. After he was relieved the next morning, he took her to the main gate, explained what had happened, and she had no trouble getting home. Among the things they recorded were seeing dive bombers dropping their bombs on Hickam Field. (They lived with Miriam's parents at Ft. Kamehomeha, which overlooked the approach to runway #5.) They saw the red meat balls on the wing tips of the planes and realized they were Japanese. They saw a B-17 and a B-18 destroyed on the ground while trying to get into the air. They saw a flight of B-17s from the mainland coming in to land, and the first one to touch down was set on fire. The others went around and flew off in every direction. They had no armament as they were in a stripped-down ferry status, and they also had very little gas, so they had to land soon. Miriam's mother insisted that Harvey drop them off at the Hickam hospital, where she knew they would be needed. Harvey went to Landing C, where boats were bringing in wounded to the hospital. He was one of many officers and enlisted men who returned in one of those boats to Ford Island. He saw the battleship *Nevada* underway, firing some 16-inch shells at the second wave of B-17s coming into Hickam. None of the B-17s were hit, but considerable damage was done in downtown Honolulu 12 miles away. There was nothing to do around the squadron hangar, so Harvey went to the airstrip and saw a PBY pilot (class 120 C) get into an SBD, start the engine and take off. That night two of Harvey's classmates from the *Enterprise* tried to get into Ford Island. One, Herbe Mengas, was shot down and killed, and the other, Dave Flynn, abandoned his attempt after getting several holes in his plane. He flew until he ran out of gas, then bailed out over the Marine base at Ewa. One other thing Harvey reported was that the water mains from Honolulu to Ford Island were ruptured. So water from swimming pools on the island had to provide all water needed, but the chlorinated water made bad coffee!

In the meantime, Wake Island was being bombed

every day. On the second day, at about the same time as on the first, 36 Bettys came over. Pilots Kliewer and Hamilton shot down one plane, A/A fire brought down another, and at least 14 planes were damaged, with 3 of these crashing into the sea. Twenty-six Bettys came on the third day, and several were shot down, including two by "Hammering Hank" Elrod. On the fourth day, Wake's defenders repelled an invasion force of one heavy cruiser, two light cruisers, six destroyers, and two troop transports. The landing troops did not make it to shore, as all of the invasion ships were damaged, and two destroyers were sunk.

The forces on Guam held out only two days before the Navy and Marine fighters there had to surrender. In the Philippines, Capt. F. D. Wagner, Commanding Patrol Wing 10, dispersed his 28 PBYs to secret hiding places on inland lakes so that they would not be destroyed on the ground like the planes in Hawaii. Although this saved the planes initially, fewer than half of them were left to fight after the first four days, and only a handful were able to retreat to the Dutch East Indies to join the Dutch, Australian, and Army Air forces after the first two weeks.

In the Atlantic, Ensign Danny Liane, a Pensacola classmate, was the first cadet trained pilot to get credit for sinking a German U-Boat. He was flying a seaplane from the USS *Brooklyn* off the coast of Martinique when he spotted the sub on the surface; he dropped his depth charges as it was diving. He received a letter of commendation from Secretary of the Navy Frank Knox. It was a little early in the war for medals.

Although Marine gunners and aviators had repelled the first invasion attempt, on Wake they knew there would be others, and it was only a matter of time until one would be successful. They endured bombing attacks every day, and some days as many as three. Their fighter force was down to two planes, mainly due to the heroic efforts of Lt. John Kinney, when on 20 December Ensigns J. J. Murphy and Howard Ady arrived from Midway with news of reenforcements that were to come from the *Saratoga*. They left with Major Bayler aboard, less than two hours before waves of carrier planes (Zeroes and Vals) began making their attacks on 21 December. The next day Captain Freuler and Lieutenant Davidson intercepted 33 Vals and 6 Zeroes. Freuler shot down two planes, but was so close to the second one that he damaged his plane and had to crash-land just before he passed out from bullet wounds. Davidson was shot down at sea by a Zero, and that was the end of VMF-211 aircraft. Major Putnam mustered his survivors, less than twenty, and volunteered their services as infantry. They gave an out-

A PBY-4, typical of the Catalinas of World War II, in full wartime star-and-bar markings.

standing account of themselves, with "Hammerin Hank" Elrod piling up dead Japanese in a semi-circle until he was shot in the back and killed by a sniper. He died with the pin still in a grenade he was holding in his hand. He received the Medal of Honor, posthumously. There were many such heroics before Commander Cunningham decided on 23 December 1941 it was time to put on his dress blue uniform and surrender unconditionally the about 400 Naval and Marine personnel and 1,000 civilians. Instead of making all of these prisoners of war, the Japanese kept about 200 civilians to work as slaves, in violation of the Geneva convention. They were never heard from again.

The *Saratoga* did not get to deliver reinforcements to Wake. When she learned that Japanese carrier planes had attacked, it was decided that such a delivery was too risky. We had only three carriers in the Pacific at the time, and no battleships, so we could not risk any more losses at this stage of the war, even if the forces on Wake were doomed to fall. The carrier returned with the fighter planes to Pearl Harbor.

Being in a war zone, with black-out conditions on the island every night, brought our social activities to a standstill, particularly where women were concerned. All of the lady tourists and military dependents were being evacuated as quickly as possible. We flew war patrols on a regular basis every third day; otherwise we had no official duties except as Duty Officer about once a month. The Officer's Club across from the BOQ was open every night, but no food was served. Our social activity was limited to drinking and gambling. The bar was in the cellar, and the walls were lined with slot machines. Drinks were still being served at the pre-war price of 15¢, and the Club had made a game room out of one of the storage rooms. There was one large poker table and several bridge tables. There were Acey-Ducey, backgammon, and cribbage boards available, and the room was usually filled with gamblers and cigarette and cigar smoke. Sometimes several of us who played musical instruments would have a black-out jam session in the ballroom upstairs. I played an alto sax; "Muck" Muckenthaler played trumpet; Stan Kent, the station Supply Officer, played clarinet; "Fuzzy" Walters played piano; and one of the other station officers played bass fiddle. We often played until the wee hours of the morning, but there were no dancers and very few listeners. If I wasn't playing music, I was playing poker, red-dog, or backgammon.

The poker games broke up every night, but I was involved in a red-dog game that lasted for three days. I finally came out ahead when I bet $750 (my car registration) and won. I held two aces, a king, and a jack, with all suits covered, to beat the dealer's card.

It was just about Christmas when I learned that my parents and Jerry Ringham had received letters I wrote on 10 December. On that date I had written lengthy letters to both setting forth everything that happened on 7 December. My parents preserved the letter, and I still have it today. It had been mailed before the censors got active and was untouched by them. In Jerry's letter, she explained that she had stayed in Los Angeles for a week trying to get to Hawaii, and she had finally returned to Vincennes safely by train. Her plan was to try to get back into Medical School in Louisville for the second semester, which was to begin in mid-January. It made my Christmas when I got letters from my sister, my grandmother, my parents, and Jerry.

By the end of December, the entire squadrons of VP-51 from Norfolk and VP-72 from Quonset Point, RI, were in Kaneohe flying combat patrols. VP-22 and VP-24 had been deployed to the Dutch East Indies to replenish the planes lost in Patrol Wing 10, and VP-44 had been sent to Ford Island to replace them. All of the Kaneohe and Ford Island squadrons were up to full plane strength, and we all expected to be deployed to the South Pacific early in the year. Flight crews were flying 1,500-mile searches with four 250-lb depth charges which were to be dropped in case of enemy submarine contacts. VP-12 lost the first plane and crew on such a patrol when Ensign Sosnoski failed to return. VP-13 with four-engine Coronados from North Island began flying to Kaneohe about this time and were soon to make bombing raids on Wake and other islands held by the Japanese in early 1942. New planes from San Diego had replaced all planes lost at Ford Island and Kaneohe on 7 December, and most of the original squadrons had been deployed to the South Pacific, or were making plans to do so by the end of January 1942. All military dependents at Kaneohe had been evacuated by this time, and because the BOQs could not accommodate all of the pilots, the evacuated quarters were made available to some of them. About mid-February, five Ensigns including myself (George Poulos, Hubert Reese, Willie Howe, "Whiskey" Willis, and me) moved into a three-bedroom, two-bath house. We planned and cooked our own meals, but had a mess steward by the name of Rogers who came in mornings and evenings to wash dishes, clean house, and make up beds. It was much more fun living in a house with a good group than living in a room at the BOQ. We established one bedroom as the ready room, where pilots who had to fly the next day could get a good night's sleep, and of course we established an open bar with each man providing his favorite ingredients. Also, this type of living provided us an opportunity to have house guests. Every Friday afternoon, the boys who were not flying would flip coins to see who would go to Honolulu to pick up girls for the weekend. Some of these girls could cook, and quite often the weekend meals were better than we had during the week. Before we moved into the house, the windows had been completely blacked out with paint, so we could have lights inside. We settled down to flying 12-hour patrols every third day and taking our fun where we could find it.

In the early days of the war one of our pilots in VP-11, George Poulos, was assigned as one of three planes to operate with the tender *McFarland* (DD237/AVD14) (a converted World War I four-stacker destroyer) anchored in Hilo Bay. These planes made dawn and evening patrols around the big island looking for evidence of Japanese invasion forces from submarines. On 26 December, after George had completed his morning patrol, he was ordered to proceed to the south tip of the island and set up an expanding search for two planes in VP-51 which were down at sea. They had become lost during their flight from San Diego and had to land in the ocean after exhausting their fuel. George had found nothing before dark and thus decided to calculate where the planes might be if they had applied reverse magnetic variation in their navigation and had flown in that direction until they ran out of gas. He proceeded to that point and located the first plane within a few miles of it, about an hour after dark. He found out from the blinking Aldis lamp in the plane on the water that the other plane had continued on course, and George found it about 20 miles from the first. He radioed Pearl Harbor the location of the downed aircraft and indicated he was returning to Hilo Bay. Pearl acknowledged his message, but ordered him to remain on site until the *McFarland* arrived before returning to Pearl. George reduced his power setting on each engine until it coughed and was hanging in the air at about 70 knots flying between the two downed planes when he saw the searchlight of the *McFarland* on the horizon. When the tender got close enough to communicate by blinker with the first plane, George headed for Pearl Harbor. He had 440 miles to go, and even with minimum power settings and a favorable wind, it would be close. He was asked to make position reports every half hour and report when he was 30 miles from Koko Point (one of the wartime island approach points). When he made his 30-mile report, he was told to proceed directly to Pearl. When he got to Diamond Head, floodlights lit up Pearl Harbor (the first time since 7 December), and George landed in the Dry-Dock Channel. He ran out of gas while taxiing, and had to be towed to the ramp.

During the first week of war, Army, Navy, and Marine Commanders had declared Hawaii a war zone and ordered the evacuation of all military dependents. The Matson liners and other cruise ships were converted to war transports, bringing food and supplies to the islands and taking back dependents. One of the evacuation stories I remember about Kaneohe concerns the wife of an Executive Officer of one of the squadrons. George the Exec was a tennis buff, but his wife cared little for tennis, and while George was on the courts his wife spent much of this time courting an Ensign George in the squadron. George the Exec heard rumors about this, and decided George the Ensign had to go. So the Ensign requested orders before the war started. These came in two weeks after 7 December, and he was evacuated on the same ship with George the Exec's wife! Almost all dependents had left by February 1942.

Before the end of the first month of war, VP-11 suffered its first plane loss since 7 December. LCdr. F. R. "Snooks" Jones (acting commanding officer) was returning from a long-range war patrol at about 2000 (two hours after dark). The sky was overcast and the island of Oahu was completely blacked out. While making a landing at Kaneohe, Lt. Commander Jones stalled the plane at about 100 feet altitude and crashed into the bay. He and four crewmen were killed. Chief Naval Aviation Pilot "Ace" Holbrook, the navigator, and three other crewmen were injured but were rescued and survived. Shortly after this accident VP-11 received a new commanding officer, LCdr. C. C. "Spud" Marcy. Lt. William "Bull" Schroeder remained our Executive Officer.

The next incident involving a VP-11 plane occurred in January 1942. Kaneohe was sending out daily long-range patrols as Patrol Wing One under the command of Cdr. M. A. MacGinnis. Ensign Carlton H. "Dopey" Clark with Charles F. Willis as co-pilot were returning from one of these 700-mile patrols on a dark overcast night with the island blacked out, and they missed Kaena Point, the island approach point on the northeastern tip of Oahu. It was the lowest spot on the eastern side of the island, and one could see the waves rolling ashore, even on a dark night. However, if navigation was off more than five miles after a 1,500-mile patrol, one could not see the waves breaking. Clark eventually exhausted his fuel without making a landfall and had to make a "dead-stick" landing in six- to eight-foot seas at midnight. He made a perfect landing under the conditions, and there was no damage to the airplane. However, it was about three days before they were located, and the *McFarland* was dispatched to the area to give them some fuel. They took on 300 gallons of gas, and Clark took off in the heavy seas and returned to Kaneohe without injury to personnel or damage to the aircraft.

The next tragedy for VP-11 came on Easter Sunday, 1942. Jack Coley, Willie Howe, "Muck" Muckenthaler and I all had long-range patrols that day. We were all airborne by 5 a.m. About two hours into the patrol, Howe discovered that his batteries were out, and he had no communications. Under such circumstances planes were ordered to return to base, but Howe could not report that he was returning. It was 9 a.m. when he returned, got another airplane, and started his patrol again. So his sector was covered about four hours later than it would have been if he had not had trouble.

About 95 percent of the time the flying weather in and around the Hawaiian islands is under visual flight rules, and most of that time the ceiling and visibility is unlimited. The weather was fine on the outgoing leg of our patrols that Sunday, but about 300 miles from Kaneohe we encountered less than instrument conditions where the ceiling was under 200 feet and visibility was less than ½ mile. I encountered these conditions at about 3 p.m. and asked the radioman to get a weather report from Kaneohe.

The answer came back, 200 feet and ½ mile, so I told my co-pilot and navigator, Ken Kemper, to give me a heading that would take us 40 miles due East of Kaena Point. I took this heading, and made corrections as Ken requested. According to his drift readings and the compass heading corrections he was giving me, we had a southwest wind of about 35 knots. This was twice the velocity of the prevailing trade winds which blew from the east northeast, so I knew the weather near Oahu was most unusual. When we reached a point we considered to be 40 miles east of Kaena Point, I circled and requested the latest Kaneohe weather. The ceiling was still 200 feet, but visibility had dropped to less than ¼ mile. The wind was 35 knots from the Southwest, as I dropped our alitutde to 100 feet and set our antenna loop direction finder so that we could "home" in on the Honolulu broadcast station. I figured this heading would bring us close to Kaneohe Bay, and I would hold it for 20 minutes, or until there was no wind. I knew I was heading directly toward the sheer mountain cliffs that lined the eastern shores of Oahu, and I also knew that when I reached a no-wind condition, I was getting close in the lee of the 3,000-foot cliffs. As it turned out, I reached the no-wind condition in exactly 20 minutes, so I turned to a heading that would parallel the island chain and was staring at the 800-foot peninsula that extends from the entrance to Kaneohe Bay, in less than one minute after I completed the turn. I was the last plane to get into Kaneohe that day, as visibility was almost zero and the ceiling was under 100 feet. It was after 1700 by the time the beaching crew got the plane on the ramp, and we had been in the air more than 12 hours. I asked the beaching crew if we had any other planes out, and was told that the other two VP-11 planes were still in the air. I also learned that Muck had wisely turned around when he found that Kaneohe and Pearl Harbor were below minimums and had radioed that he was proceeding to Nawilliwilli (a lagoon near the island of Niihau) to land and anchor for the night. I asked Ken Kemper to turn in my reports and went directly to the communications tower to "sweat" in the two planes, because I knew there would be incidents in every airplane that was in the air around Ohau that day. By the time I reached the tower, Kaneohe was "socked-in," with zero visibility. I heard Jack Coley asking Ford Island tower for weather, and they reported 200 feet and ¼-mile visibility. Jack responded by saying he had made a landfall at Barber's Point and was proceeding to Pearl Harbor Drydock Channel to land. Jack landed at 1745, and it was almost completely dark by the time the beaching crew had him on the ramp. In the meantime, I discovered that the other VP-11 plane out, commanded by Willie Howe, had begun his patrol four hours later than the other Kaneohe planes, as he had returned for another plane when he realized the batteries were out and he had no communications. In the tower we could hear Willie requesting weather from Ford Island and Kaneohe, but each time both stations reported no ceiling and no visibility. About 2100 Willie reported to our tower that he was in the clear, in the moonlight at 2,000 feet over the island of Molakai. This was very encouraging, because we figured if he knew where he was and he was in the clear, he could always make an open-sea landing in the lee of one of the islands. About 2130 we heard the Ford Island tower indicating their ceiling was 200 feet and visibility ¼-mile, and we heard Willie respond that he was coming in to land. This was the last we heard from him. About an hour later, Pearl Harbor got a report that civilians living near Makapuu Point had heard a plane crash. When Captain Marcy got the word, he sent a search party, led by Lt. Joe Cobb, to investigate. Once there they found the plane had crashed into the side of the mountain, and there were no survivors. This news hit us right in the heart of our living quarters. One of the five was gone, and I was the officer chosen by Willie to inventory his effects in case of death and ship them to Nashville. This turned out to be my saddest duty during my 20-year career as a Naval Officer. I sent Mrs. Howe a dozen red roses every Easter until she died. Willie and his crew were not the only casualties that Easter Sunday, as a B-17 and two Army P-40s crashed due to bad weather.

The next day, I was busy getting Willie's effects together, and Willis had a combat patrol. I didn't feel like eating anything all day, and went to the Officer's Club to play some poker before Willis returned. It was 11 p.m. when I returned to the house. George and Hubert were in the ready room getting some sleep before their patrols the next morning, but I could not find Willis. I finally went to the bathroom, and found him in a tub full of water. He had consumed a bottle of rum and was making a few bubbles before he sank peacefully under water. I pulled him out of the tub and began explaining that the loss of another pilot in the house was not going to bring Willie back. I then learned that he had been blinded by searchlights as he approached Kaneohe and he almost crashed as he landed in the bay. We were almost down to three. However, before the week was out, our good friend Bob Corlett had moved in with us, and we were five again.

During the first few months of war in the Pacific, there was very little good news about our efforts against the Japanese until 18 April 1942. On this date, Admiral Halsey's Task Force 16, with the carriers *Hornet* and *Enterprise,* reached a point 640 miles from Tokyo before they were detected. Here they launched Jimmy Doolittle and his sixteen B-25 bombers to bomb Tokyo, Yokohama, Yokasuka, Kobe, Osaka, and Nagoya. Each plane carried 2,000 lbs of bombs, and they were all to land on runways in Free China after they dropped on targets. The original plan called for them to launch about 200 miles closer to their targets, and most of the planes would have reached the designated landing fields if they had been able to carry out the plan. As it turned out, only one plane was able to make a routine landing on a runway, and that was at Vladivostok, Russia. Most of the planes delivered their bombs on targets and escaped with little anti-aircraft fire, because the surprise was so great. Like the attack on Hawaii, no one thought a bombing attack on Japanese cities was possible. Except for the plane that landed in Russia, the other 15 planes crashed. Some of them crashed at sea, some made belly landings in rice paddies; but most flew until they were almost out of fuel and bailed out close to free China. And as incredible as it seems, most of these brave men survived this one-way mission, exceeding their own expectations. The destruction wrought by their bombs on the Japanese homeland bolstered morale of U.S. forces everywhere.

On 6 May 1942 our forces in the Pacific reached the low point in the war when we surrendered to the Japanese at Corregidor. At the same time, a Japanese invasion force was heading for Port Moresby. Our Task Forces in the Coral Sea with the carriers *Yorktown* and *Lexington* had been ordered by Admiral Nimitz to stop the invasion forces at all costs. This they did when our dive-bomber pilots sank the carrier *Shoho,* severely damaged the *Shokaku,* and our fighter pilots destroyed many of the planes of the *Zuikaku.* This was the first battle in history where strong carrier forces were opposed yet none of the ships were ever close enough to see each other. We lost the tanker *Neosho,* the destroyer *Sims,* and the carrier *Lexington,* as well as many pilots including Carl Rinehart's brother, Clark. But the Japanese losses were so great that they decided they could not control the air around Port Moresby, and they called off the invasion.

With the Port Moresby invasion postponed, Admiral Yamamoto decided to gather all of his forces for an all-out invasion of Midway. Our intelligence officers had broken the Japanese code, and Admiral Nimitz was aware of the fact that an attack on Midway was imminent. As early as the middle of May, he began sending aircraft reinforcements to that island. The Kaneohe squadrons, including VP-11, began checking out in new PBY-5As, which were amphibious and could operate from land or water. Our squadron got six of these planes, and flew them to a gravel landing strip at Barking Sands, Kauai, on 19 May. This allowed our 700-mile patrols to get 100 miles closer to Midway. The other six planes in our squadron flew to Johnston Island to operate from the water and thus to bring their 700-mile patrols closer to Midway. There were 18 PBY-5As at Kauai, and 18 PBY-5s at Johnston Island. All of VP-23, under the command of Massey Hughes, went to Midway in the last week of May, as well as planes from VP-44, 51, and 72. With these planes and General Tinker's B-17s from the Seventh Air Force, long-range

A PBY-5A amphibian in full flight with radar over the cockpit and fully extended fin and rudder. (General Dynamics — Convair)

search patrols flying every day covered sectors of 180 degrees to 700 miles west of Midway (700 miles was the magic distance Fleet Commanders decided would allow 24 hours of safety from enemy attack). By the end of the day on 3 June, there was a total of 120 planes of all types, including 4 B-26s which were rigged to carry torpedoes on the island. Our forces at sea included the carriers *Yorktown* (repaired from Coral Sea damage), *Hornet,* and *Enterprise,* accompanied by 8 cruisers and 14 destroyers. The Japanese had more than 200 ships headed for Midway, including 8 carriers, 11 battleships, 22 cruisers, and 65 destroyers, as well as 20 submarines. By the end of 5 June they had lost 4 carriers, about 325 airplanes, and many of their finest pilots. The U.S. lost about 150 planes (35 of 41 torpedo planes), but many of the pilots and crewmen were picked up by destroyers or PBYs. The PBYs were christened "Dumbos" after they had picked up 27 people including a PBY crew that was shot down. (Dumbo was the flying pink elephant in a new feature-length Walt Disney cartoon.) Lt. Norman Brady, a Pensacola classmate, flew out over 450 miles on 6 June to locate and rescue the crew of a PBY that had been shot down on the 4th. Norm said he got them all safely back to Midway, except for the Chief Aviation Machinist Mate who died enroute. Ensign George Gay, the only survivor of Torpedo Squadron 8, was picked up by a PBY. He gives a good account of his ordeal in his book, *Sole Survivor*. Jack Poteat, a VP-72 pilot, was returning from a 700-mile patrol on the 4th, more than 500 miles from Midway, when a B-17 flew abreast of him and asked by blinking light for a course and distance to Midway. Jack blinked in answer, "follow me." He got the B-17 to the island and it landed safely, but it ran out of gas before it got to the flight line. Many of the B-17 pilots had trouble navigating at sea and had to "ditch" their planes when they ran out of fuel. Three of the B-26 pilots made it back after dropping their torpedoes. All in all, Army, Navy, and Marine pilots turned out to be heroes during the battle that proved to be a great victory for our forces and the turning point of the Pacific war.

On 16 June 1942 we ran regular patrols from Kauai, but we returned to Kaneohe and learned that the entire squadron would be deployed to the South Pacific beginning 1 July and that our plane strength would be increased to 15. The planes were to leave in three-plane sections each day. I was scheduled to depart in the last section, and the Executive Officer, "Bull" Schroeder, was to fly with me. I had been advertising my 1940 Plymouth for sale in the Sunday papers for a month, and I had no calls, even though I was asking only $500. Civilians could not get gasoline, and military people did not have much use for autos. In desperation, I took the car to a used car dealer in downtown Honolulu on 2 July. I asked for the man in charge, and when I found him I asked what he would give for the car. He drove it around the block, stepped out of the car, and said, "I'll give you $750!" I was so happy, I took a cab back to Kaneohe. The fellows in the red-dog game had been

A USAAF Martin B-26 fitted with a dummy torpedo. (USAF)

about right in allowing me $750 for the registration in my bet against the dealer.

We were all ready to leave early in the morning of the 5th, but George Poulos could not get one of his engines started, and Bull Schroeder decided it should be changed. This required more than one day's work, so we rescheduled departure for the 7th. George and I had been given a big farewell party on the 4th with fireworks and firewater, and when we called the girls and told them about the delay, they insisted on having another going away party on the 6th. George and I accompanied the girls to the grocery to get the "fixins," but we had to get into our deployment ration as far as the whiskey was concerned (each pilot was allowed one gallon when deployed). We flipped, I lost the toss, and was carrying my gallon in one arm and a bag of groceries in the other, when the screen door banged against the whiskey arm and we had Ten High all over the living room. I cleaned it up while George went to the house and got his gallon. We drank most of it, and this made for a very happy departure.

Before deployment, each pilot in the squadron was allowed to have a cruise box constructed by the Public Works Department of the Station. The boxes were made of good heavy plywood and were designed to store safely all of the pilot's clothing and personal effects, except khaki uniforms and flight gear, which included a .45 cal pistol and jungle knife. Russell Delaney was the officer left behind to assure that all cruise boxes were aboard a Supply ship that would take them to a safe storage area in the deployment zone. As luck would have it, my cruise box never did arrive, and I assumed that it had been aboard one of our ships that was sunk. I did not learn until January 1944 that my box did not leave Kaneohe. It was sitting in the damp basement of the BOQ, until the Officer-in-Charge shipped it to me in Jacksonville. About the only thing I salvaged was my sword.

Our three-plane section took off for Palmyra Island at dawn on 7 July, flew to Canton Island on the 8th, and arrived in Suva, the capital city of Fiji, on the 9th. There we joined six other plane crews and assumed responsibility for long-range search and anti-submarine patrols for the Fiji Islands. Our planes relieved the old New Zealand four-engined Singapores, who had this responsibility previously, and the U.S. Army relieved New Zealand Army forces, who were responsible for the security of the islands. Six of our plane crews, led by Lt. Joe Cobb, had proceeded to the New Hebrides to fly war patrols from the aircraft-tender *Curtiss*. Our nine crews were taking our meals and sleeping at the Grand Hotel, which was the best hotel in town. There was double occupancy in each room, and curtains served as doors, but we did have mosquito nets on the beds. The weather was dreadful flying out of Suva, so Captain Marcy decided we should find another base of operations, preferably one that allowed us to begin our patrols farther westward toward the Japanese. We found a perfect spot in Suwena Bay on the western side of the island near the second largest city, Lautoka. Bull Schroeder and I were designated to establish a base there as soon as possible. With the help of a shipload of SeaBees, we built a base at Suwena in ten days, all without funds or appropriations. We had mooring buoys in the bay, a concrete ramp and apron where we could get planes out of the water, and even fashioned a chain hoist from a coconut tree so that we could lift an engine from its mount to work on it. We established a general mess and living quarters for all hands, which included mosquito nets on all beds and bunks. During the building period, Bull and I stayed at the Lautoka Hotel and were instrumental in getting information to Army intelligence which resulted in the capture of two women who were operating as German spies supplying information to the Japanese.

About the time we reached the South Pacific, the AVGs as a group were disbanding. Seventeen of these gentlemen decided to stay in China to work for the CNAC (Chinese National Aviation Corporation) as senior captains flying American planes over the Hump (Himalayan route). These pilots included ex-AvCads Bill Bartling, Link Laughlin, Joe Rosbert, and Dick Rossi. At least two pilots trained as Navy cadets decided to join the Army Air Force at this time, and some others elected to do so later. David Lee "Tex" Hill was advanced to the temporary rank of Brigadier General before he had been in the Air Force two years, and was one of their youngest officers to wear a star. (He made BG before I was made a LCDR in the Navy!) After a highly successful career, he was retired as a Colonel in the Air Force. Jim Howard, the other cadet-trained AVG who joined the Air Force, was sent to the European theater flying P-51 fighters over Germany. One day in January 1944, he shot down three German fighter planes before his ammunition was expended. For this he received the Medal of Honor, the only fighter pilot to be so rewarded in Europe. After the war he remained in the Air Force Reserve and was retired as a Brigadier General. Other Flying Tigers like Greg "Pappy" Boyington went back to the service that trained them. Pappy was another Medal of Honor winner with the Marine Corps, was a prisoner of war on the Japanese mainland, and had a successful career to retire as a Colonel, USMC. Most of the AVGs were aces when the group disbanded, and almost without exception all enjoyed successful careers in aviation with the Air Force, the Navy, or the Marine Corps, and many became Senior Airline Captains. A few became entrepreneurs when they started the Flying Tiger Air Freight airline.

Although we had built a fine little Patrol Plane Base at Suwena Bay in ten days, we only had the use of it for three weeks. On 24 August we all flew to the New Hebrides to run patrols and use the seaplane tender *Curtiss*. This was the only time since the first of July, that we had all squadron planes in the same area. On the 25th Lt. Carl Clark took off from Espiritu Santo (code name, Buttons) in the New Hebrides at dawn and opened his sealed orders. He was to proceed to the island of Ndeni in the Santa Cruz Islands, capture it, and await further orders. Jack Clinton, his navigator, got out the maps and learned that the island was about 400 miles NNE of Buttons. Clark circled the island at low altitude and could find no sign of life, so he decided to land in Graciosa Bay, which was about two miles long. It was protected on three sides before extending into the open ocean to the north. Clark and his crew chief went ashore in a rubber life raft and gave Clinton and crew orders to take off if it looked like they were to be captured. They were back with the raft within an hour and had found only one lonesome native who lived in a grass shack. He could neither speak nor understand English. The crew settled down to await further orders. Shortly after noon they spotted a ship lying off the entrance to the bay and thought it was surely Japanese. But as it came on into the bay, they could see it was an old four-stacker U.S. destroyer, which was to act as a tender for the PBYs that would be coming in later in the afternoon. It was the *Ballard* (AVD 10), and it would be joined two days later by the *Mackinac*. Between them they handled about 24 planes, of which 15 were from VP-11. From Ndeni, our patrol sectors were about 350 miles closer to Guadalcanal, and our patrols were making daily contact with Japanese units. On the 27th, one of our planes from Ndeni made contact with a Japanese battle fleet. There were no carriers in the group, but the plane that made contact had done so at the end of his 700 miles and did not have enough fuel to track the fleet for any length of time. George Poulos was assigned to take off at noon, make contact with the fleet before dark, track it until midnight, and return at daylight. He found the fleet about 150 miles northeast of where the last contact had been made, and after tracking the ships until midnight, decided they were heading for Truk and would be no threat to our forces in the Solomons. Shortly before daylight, George knew he was getting close to Ndeni when he could smell the fumes from an active volcano that was 80 miles north of it. He came on in after an 18.4-hour flight with 30 gallons of gas left, and landed at first light, as the regular patrols were taking off on their search missions. Our patrols at Ndeni were then stepped up to two days out of three.

On 31 August, I was told to report to the Commanding Officer's stateroom. When I got there the Captain of the *Ballard* handed me sealed orders, which I was to open after takeoff. It was about 10 a.m. when I got off and opened the orders. They indicated that the

The much-respected Japanese Zero single-seat fighter coming into land. It showed that the Japanese had fully grasped the technological revolution. But they failed to up-rate it or supersede it in time during the war. (USAF)

carrier *Saratoga* had been torpedoed in the vicinity, and I was to begin an expanding search at a specific longitude and latitude until I found the ship and ask if I could render assistance or fly anti-submarine patrol in the area. I found the ship about noon. She was trailing oil, with a ten degree list to port, but she was recovering some planes that had been in the air, and she was making about 15 knots. I remained in the area for about an hour, and after blinking several messages back and forth was told to return to base. She thanked us for our offer to help, but indicated she could maintain 15 knots and was headed for Pearl with her destroyer screen.

I flew a regular patrol the next day and spotted a four-engined Japanese seaplane, but I was not close enough to attack him. On 3 September, Lt. F. G. Riley, a VP-23 pilot operating with VP-11 planes out of Ndeni, crossed paths with the same type of plane I had spotted and shot it down. One of our sectors was spotting a Japanese patrol plane at about the same time and place each day, so on 6 September Clark and Willis decided to run their patrol backward in order to converge on him. They also ran their patrols that day at 6,000 feet so that they would have plenty of speed to attack, because they knew the Japanese could run away from our planes in level flight. Willis attacked the enemy first, then ''Dopey'' got him from the other side. But Willis was shot down, and his chief mech Franovich (we called them plane captains) was killed. Clark continued his attack, killed the tail gunner and side gunner, and knocked out one of the plane's engines. At this time the Japanese thought he was about done for and tried to ram Dopey's plane, but Dopey turned into him and thought his bow gunner would finish the Japanese plane. But the bow guns were jammed and the Japanese got away, though he may not have made it back to base. Clark's starboard wing float was shot off, so he decided he could not land and pick up Willis' crew who had signalled from the downed plane that they were OK. Clark kept a very accurate record of Willis' bearing and distance from Ndeni, but it was three days before they were spotted in the rough seas in their life rafts. They were picked up by one of our regular patrol planes on 10 September. Meanwhile, Clark and his crew came into Ndeni two hours after dark on the 6th, and both tenders turned on their searchlights. Little did we now that a Japanese submarine was on the surface at the entrance to the bay taking bearings on the two tenders. About midnight, after everything had quieted down and everyone had hit the sack, the submarine started lobbing three-inch shells into the bay. Neither the tenders nor the planes moored in the bay were hit, but a lot of dishes in the wardrooms of the two tenders were broken when their five-inch guns started barking back.

On the day Willis and his crew were picked up, I found a four-engined Japanese and made one good run for my bow and port waist gunners, but we did not shoot him down. On this same day, Clark ran into some Japanese Zeroes on floats at the end of his sector, and they shot him down. They shot both of his engines out, and he had to make another dead-stick landing in the open ocean in heavy seas, with several Zeroes firing at him all the way down. He made a perfect landing, but his plane had so many holes in it that it sank. Soon after, the crew was picked up by a destroyer and taken to the ship that had launched the fighters by catapult. The ship, one of the largest and fastest ships in the world, could launch up to eight Zeroes on catapults, but had to recover them one at a time by cranes. Clark and his crew were prisoners on the Japanese mainland for just over three years.

Also during this week, George Enloe sorely damaged his plane during takeoff in the huge swells with a cross-wind one morning, but continued his patrol. This was quite risky because he knew his plane would sink if he had to land at sea. All crews aboard the tenders knew George had damaged his plane on takeoff, and we were all on deck to observe his landing. The landing was smooth enough, but the plane kept getting lower in the water as he was taxiing toward a mooring buoy. The crew finally had to break out the life rafts and abandon the sinking plane, but the engines were so hot that the propellers kept turning in a mass of steam until the plane was about under water.

We had lost three planes in a week, and we had been flying an average of about seven hours per day. Some of the pilots and crew members had been grounded by flight surgeons for various reasons, and our Skipper, "Spud" Marcy, decided it was time to give his crews some R & R. Luckily, we had lost only one flight crew (Clark's). When VP-91 reported to the *Curtiss* with new planes and fresh crews, Captain Marcy requested and was granted authority to rotate flight crews to Auckland, New Zealand, for R & R. On 21 September, George flew Muck Muckenthaler's and Charlie Willis' crews to Noumea, and thence to Auckland on the 22nd, to set up the rotation. The Exec, LCDR Bull Schroeder, went along with the first crews to keep everything on the up and up. One plane would fly down two new crews each week, and the crews that had the R & R would fly the plane back. Bull, Charlie, and Muck did a great job in establishing the rotation, as they acquired the names and telephone numbers of 200 unmarried lasses from the finest families in town who were eager to entertain some American fliers who had been fighting to save their country. The rotation for Bob Corlett's and my crew started on 6 October when we flew to Noumea, arriving in Auckland on the 7th. New Zealand was the most beautiful country I had ever seen. It stretches from the tropics at the northern tip of North Island to the Antarctic at the southern tip of South Island. The lucious green fields of grain, interspersed with herds of cattle and flocks of sheep grazing, was a sight to behold.

Our residence in Auckland was the Star Hotel, at the time one of the best in the city. We arrived in the late afternoon, and our first meal was dinner that night. It was the first time we had seen anything fresh since we left Hawaii. We had fresh milk, fresh steak, fresh fruits, and fresh vegetables, but not until we had been through the last half of their happy hour. The hotel served drinks from 4 p.m. until 6 p.m. each day, and you could not buy a bottle. We called the bar at 5 p.m. and told the bartender to start moving one waitress with a tray full of drinks to our room, and to have her make as many return trips as possible until he closed. He did not seem surprised, so we guessed that the fellows who had preceded us had done the same thing. We passed the word to the bartender via the waitress to start the process the next day at 4 p.m. The drinks were all scotch; they had never heard of bourbon, and you could not drink their rum. We had every window sill and every dresser, stand, and table loaded with scotch. It is hard to believe how much a few drinks and a well prepared five-course dinner with all fresh foods can change your outlook on life. To this we added a cigar course, and repaired to a lobby filled with good-looking young ladies who were waiting to go to the ballroom for some dancing. This proved to be one of our best wartime operations. We were invited to picnics, meals and parties at individual homes, sightseeing tours (although gasoline was in short supply), and we bet on the horses on two weekends. They seemed to be running backwards, as they raced clockwise like they do in England and Australia. When the plane with fresh crews for R & R arrived in Auckland on the 14th, the one that was to take us away from this wonderful world needed spare parts. This meant we had to endure this tough assignment for five more glorious days. The plane with the necessary parts arrived on the 17th, we made the required repairs the next day, and flew to Noumea on the 19th. We were back at Buttons (Espiritu Santo), aboard the *Curtiss* on the 20th, ready to take up our 1,500-mile patrols and more determined than ever to get at the Japanese. Our trip to New Zealand had really made us appreciate what we were fighting for, as we knew the New Zealanders and the Australians deserved to be saved. Three of my good friends, fighting the war from Perth, changed the lives of three Australian girls when they married them. One of them, Pensacola classmate Jack Davis, still lives with his wife Mary in Victoria Park, West Australia.

When we got back to the *Curtiss*, we had several letters waiting to be read. Although the news in them was four to six weeks old, it was most welcome. Jerry was making As and Bs in medical school, sister Marjorie was courting a young intern who was about to join the Army as a doctor, and my parents were both well. I answered their letters and told them what a great time we had in New Zealand, and that I had sent them some gifts for Christmas while I was there. I also pointed out that we were back on powdered milk, eggs, potatoes, and Australian mutton. Also, we were back in the routine of flying 12-hour patrols, with a few rescue missions in between. Although I could not say anything about it, I thought about how well off we were sleeping in clean beds in our staterooms while the poor Marines were sloshing around on a beachhead at Guadalcanal trying to capture Henderson Field for the use of our pilots. This made us more determined than ever to keep the Japanese from getting at those Marines. The "Coast Watchers" had alerted our intelligence to the fact that the Japanese fleet was establishing a major base at Tonolei Harbor on Bouganville, the northernmost island in the Solomon chain. This greatly reduced the length of their supply lines from Truk and would become a real threat to our Marines on Guadalcanal. Thus our PBYs became more aggressive. The "Coast Watchers" became our most important intelligence source during this phase of the Pacific war, and they were directly responsible for the rescue of literally hundreds of pilots who were shot down, or who for some reason could not make it back to their ships or bases. They also provided advance notice of attacking Japanese planes, including types and numbers. These fellows were super intelligent, mostly Australian, who had the ingenuity to keep powerful transmitters operating and to live by themselves off the land for long periods of time. I think it is safe to say that they saved many thousands of lives during the war. Walter Lord gives a detailed account of their operations in his book, *The Lonely Vigil.*

On 22 October 1942, I had my first 12-hour patrol after the R & R. This was the same day that word came about a large carrier force being assembled at Tonolei Harbor. Vice Admiral Fitch, the SouthWestPac Commander, decided to do something about it and he chose three planes from our squadron. George Poulos, Jack Coley, and Charlie Willis with full plane crews and three torpedoes were to plan and execute the mission. Each plane also carried two 500-lb bombs, as they took off at dusk to make their attack at 2 a.m. on targets 900 miles away. Jack Coley led the flight as they flew the last 150 miles in tight formation at less than 50 feet of altitude to avoid detection. Their navigation was perfect, as they flew right into the harbor entrance and had to pull up to avoid hitting a destroyer that was guarding it. Here they broke formation, and each pilot selected a capital ship. They dropped their torpedoes from 300 yards out, then released their bombs as they pulled up to avoid hitting the targets. The crews verified the fact that each of the pilots had scored hits, and the Japanese learned that their sanctuary was not safe from the PBY. The pilots regrouped in loose formation at 1,000 feet and flew to Tulagi, where they took on enough fuel to get back to Buttons. The attack had been a complete surprise, and had been superbly executed without a single anti-aircraft shot being fired at them. The damage could not accurately be assessed, but it was enough to drive the carrier fleet to sea, where they were located and tracked by our PBYs, so our carrier pilots could get at them. This developed into the Battle of Santa Cruz, which resulted in severe losses to both sides. Though we lost the carrier *Hornet* and had several cruisers and destroyers put out of commission, the battle prevented the Japanese from mounting a major invasion force to retake Guadalcanal and Henderson Field. George Poulos not only helped drive the carrier force from Tonolei Harbor, but he located four carriers and their escorts at the start of the battle on 26 October. The Japanese carriers had already launched their planes to attack our Task Force, but our planes were awaiting the contact report and position of the enemy fleet that Poulos provided. After George made his report, he was attacked by four Zeroes on combat air patrol, but he got down between the huge ocean swells and avoided machine gun bursts from any of the four planes. This same day, George and his crew got close enough to see a submerging Japanese submarine making a swirl so that its bi-wing seaplane could take off. Several enemy submarines carried these knockdown seaplanes, but we were not aware of this capability until George spotted it.

On 29 October, while on a regular search patrol, I was flying in some low-hanging clouds at about 200 feet altitude 400 miles northwest of Buttons. When I broke out of the clouds, I saw a Japanese submarine on the surface directly ahead of us. I made my full power run and dropped my four 250-lb depth charges just about amidship as he was submerging. I then pulled up to let out antenna wire, so that we could make our contact report. We remained in the area circling the huge oil slick and debris for ten minutes, then sent an amplifying message reporting the kill.

When the PBYs were not on their primary mission of searching areas to the north and west of the New Hebrides and Santa Cruz Islands, they were sent on specific rescue flights with orders to pick up downed crews or pilots. There were many pilots in the water after the Santa Cruz battle, and most of them were carrier pilots who were picked up by PBYs. Rescued pilots gained a new respect for the lumbering P-boats. There were many cadet-trained heroes of the battle, including Lt. Stanley "Swede" Vejtasa who shot down two Vals over the *Hornet,* and six Kate torpedo planes which were making runs on the *Enterprise*. His action against the torpedo planes certainly saved the *Enterprise* from receiving more serious damage. After this battle, we had only one operational carrier in the South Pacific, and that was the *Saratoga*. Realizing this, Admiral Fitch began to use the PBYs for long-range bombing and torpedo strikes, while using Army B-17s for high-altitude long-range search planes. On 13 November, Lieutenants Coley, Corlett, and myself were called to Admiral Fitch's quarters for a briefing. An Army B-17 had sent a contact reporting twelve capital ships 900 miles northwest of Guadalcanal on a southeast heading, but there were no amplifying reports to indicate if the ships were carriers, battleships, or cruisers. The plan was to have us take off about 1 p.m., fly to Tulagi, land, and refuel at dusk, then take off and deliver our three torpedoes on the most lucrative of the twelve targets. There was one condition that could cancel the mission. If there was a Japanese attack on Henderson Field at the time we arrived at Tulagi, we were ordered to return to base with our torpedoes. At Tulagi we would have been "sitting ducks" for the Zeroes protecting the bombers at Henderson, just ten miles away. As that happened to be the case, there was nothing to do but follow orders and return to Buttons. We not only saved the PBYs, but we saved the torpedoes. Coley and Corlett elected to land among the 75 ships at anchor, which were unlighted, but I chose to delay my landing until dawn. I was flying at 5,000 feet between Tulagi and Buttons for the rest of the night. I witnessed the awesome sight of large ship engagements at night just north of Tulagi that morning. Although we did not know it at the time, the twelve capital ships turned out to be transports loaded with about 6,000 troops on each ship, and they were going to drive our Marines off the island and retake Henderson. The Cactus (code name for the Guadalcanal operations) Air Force on Henderson had other ideas. These pilots were a mixture of Navy and Marine pilots, which included cadet-trained Joe Foss and Marion Carl for the Marines, and 90 percent of the torpedo and dive-bomber pilots of Air Group Ten from the *Enterprise* who were cadet-trained. This force also included most of the scouting pilots of VS-10, as well as a large number of fighter pilots from VF-10, including Swede Vejtasa who had shot down the eight planes one day in the Battle of Santa Cruz. Once the contact reports of the PBYs accurately described the location and description of the invasion force, the pilots from Henderson began making their attacks with torpedoes and 500- to 1,000-lb bombs. The *Enterprise* retreated to the south with about a dozen fighters to provide combat air patrol. Admiral Fitch and his staff gave a great sigh of relief when they learned the twelve capital ships were all transports, and they had very little air support, with only four cruisers and six destroyers as surface support. The Cactus Air Force pounded the transports unmercifully with bombs and torpedoes until they were all sunk, and Guadalcanal and Henderson Field were safe from invasion at last.

From the time most pilots started flying, they had their good luck charms, and as they got into the war and closer to the enemy these charms seemed to mean more and more. Some had a rabbit's foot, some had lucky coins, others had rosaries, and still others had musical instruments, or even cross bones. (I had an alto sax.) At times I broke out my sax for a "Sing Along" on the *Ballard, MacFarland, Mackinac,* and *Curtiss*. The charms had little to do with what fate had in store for the pilots who treasured them, but like the American Express cards of today, they would not take off without them. As our squadron plane crews began making daily contacts with the enemy, we became a closer knit organization. If pilots or crewmen were incapacitated, others would volunteer to replace them, even if it meant flying many days in a row.

During November, I was selected for two search and rescue flights. On the first such flight I had to locate and rescue the crew of a DC-3 that had ditched 200 miles north of Buttons, in the open sea. On the second occasion, I combined a 1,500-mile search patrol with a rescue mission and picked up a carrier pilot in enemy territory who had been reported by a Coast Watcher. This search and rescue flight lasted 14 hours. During most of November and the early part of December, our patrols extended 850 miles from Espiritu Santo, and we would land that evening at Vanakoro Island. Then we would take off from Vanakoro the next morning, fly the sector in reverse, and return to Buttons.

On 7 December 1942, exactly one year after the Japanese attack, Commanding Officer Spud Marcy received dispatch authority to select six pilots for duty with the training command in Jacksonville, Florida. He selected the six who had been in the squadron for the longest time. Russell Delaney and Jack Coley had love interests in Hawaii and were lucky enough to find a PBY in Noumea that had to be ferried to Hawaii, where they married and changed the girls' maiden names to Lois Coley and Beverly Delaney. Bob Corlett, Max Ricketts, Warrant Officer Flynn, and I flew a plane to Noumea on 13 December and boarded the troopship *Washington* (acutally then *Mount Vernon,* AP-22) for the two-week trip to San Francisco. VP-11 ended its South Pacific deployment in mid-January 1943 when the pilots and crews flew back to Kaneohe for 30 days leave. They then were ordered to new squadrons.

Bob Corlett and I had started flying in the same E Base class with six other beginners in June 1939. We had soloed there and had been assigned to the same flight class at Pensacola (130 C). We had ridden in the same train Pullman car from St. Louis to Pensacola and had been assigned to the same flight wing there (Wing 3). But Bob had received his wings 6 weeks before I had, because I had been grounded for an appendectomy and recovery period of about that length of time. When Bob got his wings he had orders to report to VP-23 in Pearl Harbor, and when I got my wings I had orders to report to the same squadron. Now here we were in December 1942 on the same troopship headed for San Francisco, 15 days leave, then flight instructor duty in Jacksonville.

The *Washington* docked about 9 p.m., and we were sitting in our hotel room within an hour. For old time's sake we decided to go to the top of the Sir Frances Drake for a few drinks before turning in. I called Jerry and my parents and told them I would be leaving on the train the next morning and would be home New Year's Eve afternoon. At the top of the Drake I started talking with some girls seated nearby, and after a few drinks took one of them home in a taxi. When the cab driver got me back to the hotel, I reached for my billfold to pay him, and discovered that I had been "rolled." I asked the cab driver to take me back to the address where we had deposited the girl, but by the time we got there the people denied knowing her or anything about her. The taxi took me back to the hotel, where I awakened

Corlett to get some money to pay the driver. I then called the police to make my report to them. I lost about $900 in cash and two Navy checks, for about $300 each. By the time I had taken care of the police report the night was about over, so we checked out of the hotel and went to the railroad ticket office. Fortunately, Bob had enough cash to buy both of our rail tickets, and I gave him a personal check to cover the price of mine. We rode the same train to Burlington, Iowa, Bob's hometown, but I continued on.

When I got to Chicago, I had to go to another railroad terminal to get a C& EI train to Vincennes. I arrived with only the clothes I had on my back, a cotton khaki uniform, and a small hand bag containing three khaki shirts, some socks, and one extra pair of khaki trousers. The temperature was 15 degrees above zero, but the wind was blowing about 30 miles per hour and the chill factor must have brought the temperature down to where it felt like it was 15 below zero. When I hailed a cab outside the terminal to take me to the C & EI terminal the driver said, "Man, you had better get some clothes on." I explained that I had lost all of my uniforms out in the Pacific, so he took me to a friend of his who owned a uniform shop and lived above it. It was five o'clock in the morning, so he had to pound on the door to awaken his friend. He explained to his friend that I had lost all of my uniforms in the Pacific and was freezing to death. So his friend came down to the first floor and opened the store to measure me for a new dress blue uniform, complete with a new white hat, new white shirt, new tie, gloves, new black shoes with black socks, and a new blue overcoat. He also got his wife up to make alterations that were necessary. I told him I would have to give him a personal check, and he said that would be fine. I then told the cab driver I would take him to breakfast. After breakfast we came back to the shop to pick up the uniform, and I changed from cotton khaki to a dress blue uniform. The cab driver took me to the C & EI terminal, and I tried to give him a substantial tip, but he would only accept the fare shown on the meter. By 8 a.m. I was aboard the train on my way to Vincennes.

When I arrived in Vincennes, I was just a shadow of my former self. I had lost 50 lbs since Jerry and I had been engaged in San Diego, more than a year before. Also, my exposure to the cold winds in Chicago had worsened my cold, and I was very close to pneumonia and running a high temperature. As much as I wanted to visit with Jerry and her parents, we all thought it would be best if I went home to stay in bed until my temperature subsided. I spent New Year's Eve and the first two days of January in bed before my temperature dropped. Jerry and I spent the next ten days trying to decide if marriage was the best thing for us. She did not want to give up medicine, and I could not get out of the service until the war was over. We decided that because we could not live together and did not know when this would be possible, we would wait for a more propitious time. (It never did come.) Furthermore, the fact that I was broke after being robbed in San Francisco did nothing to bring us closer to wedding bells. Jerry decided that under the circumstances, it was foolish to remain engaged, and she returned the diamond engagement ring I had given her in San Diego. We parted good friends and have remained so.

When Corlett and I checked into NAS Jacksonville for duty as flight instructors, there were two Navy contracts awaiting our signatures. They invited us to become regular Naval Officers in the rank of Lieutenant with a date of rank, with, but after, the U.S. Naval Academy class of 1940. Although 90 percent of our contemporaries fighting the war were Reserve, we liked the way the Navy operated, and we wanted to become a part of the regular establishment. So we signed on the dotted line, and neither of us ever regretted making that decision. Coley and Delaney became regular officers, and Ricketts and Flynn were already in the regular Navy. Coley and Flynn went into PV twin-engined landplanes, while the other four from VP-11 remained in PBYs with the training command.

Jacksonville was a wonderful city and a great place for an advanced training base. Almost all of the instructors were pilots who had spent a year in the war zones, performing every type of mission that could be flown by the PBY. As instructors who had just returned from the war zones, we were well equipped to train the young Ensigns who had become naval aviators. The operations and scheduling officers did a great job in setting up fair and equitable training flights for all instructors. Roger Crowley, in Unit One, and P. K. Bloesh, in Unit Two, arranged scheduling so that instructors would fly four hours in the morning (one hour with each Ensign), be free that afternoon, then fly three hours at night (one hour with each Ensign). These same instructors would fly four hours the next afternoon (one hour with each Ensign), then be completely free the next day, and begin the same three-day routine all over. Instructors had no administrative duties, so they had a great amount of free time while instructing 100 hours per month, and morale was high. All bachelor Lieutenants and above lived in the senior BOQ, which was next door to one of the finest Officer's Clubs in the service. It had a large pool area overlooking the St. Johns River. The BOQ even had its own lounge and game room, called "Ceiling Zero." Living conditions could not have been more favorable for a 24-year-old bachelor, and I certainly enjoyed every minute of it. In such an environment it was hard to believe that we were still at war. There were many unattached young ladies in the area who enjoyed nothing more than spending an evening with a young bachelor instructor. There were many all-night parties before our days off, however, as far as romance was concerned, I had decided that I would not get married until the war was over.

Being a free bachelor, ready to go anywhere at the drop of a hat, allowed me to be available for assignment to some interesting temporary additional duty. The first was a trip to San Diego to pick up overhauled planes for the training command. I was the co-pilot for the return flight to Jacksonville, with Roger Crowley our pilot and officer-in-charge of the three-plane flight. Pilots and crews were flown to NAS North Island on Coronado Island where we were to pick up the planes. We arrived one afternoon, test-flighted the planes next day, and were scheduled for 7 a.m. takeoff the following morning. All hands were on time except our navigator, who showed up one hour late and could not have registered less than .25 on alcoholic blood tests. After we got in the air, Rog did not bother to ask him what happened; he called him up in the cockpit and told him to find out how far it was to Corpus Christi, Texas. He came back from the navigation table within a minute. He looked up at Rog with the thumb and forefinger of his right hand fully extended, and said, "It's that far Rog, as the crow flies." Rog just told the navigator to get back in the sack, and told me to take over navigational chores.

On the second TAD assignment I was flown to San Francisco on a commercial plane where I picked up a PBY and flew it to San Diego to join two other planes which were to be ferried to Jacksonville via Fort Worth, Texas. I flew from San Francisco to San Diego on 12 May 1943, but had to delay takeoff from there until 16 May. All three planes arrived at Lake Worth in the late afternoon, and two of them landed in the big lake. I could see that the planes on the lake were having a difficult time taxiing in the 35 knot wind, which was blowing right down the channel to the refueling dock. I made one pass down the channel and decided I could land there, even though it was only 600 yards long and 200 yards wide. I made a full stall landing, directly into the 35 knot wind, and I was not 100 yards from the refueling dock when the plane stopped. The fueling crew said we were the only plane that ever landed in the channel leading to the dock, but we were fueled and had the plane moored for the night before either of the planes in the lake got to the channel.

The third TDY assignment came on 24 September 1943 when I ferried a two-seat open cockpit SB2U dive-bomber to Memphis. I was checked out in the plane by Lieutenant McAboy (another instructor who was supposed to make the flight), but after two landings on his checkout flight he decided that he could not make the trip. He gave me one landing from the rear seat and one from the front, and said I was qualified, but when I taxied back to the flight line, the right tire was flat. By the time the tire was fixed, it was after noon. I took off for the three-hour flight to Mobile, but the plane was a clunker and was throwing oil all over the cockpit. I was flying at 2,000 feet, with my seat belt unbuckled ready to flip the plane on its back and bail out. I made it to Mobile in three hours, refueled, and required two gallons of oil before taking off for the one-hour flight to New Orleans. I made it to Memphis the next day, where I had a nice visit with the parents of Hubert Reese (one of the original five at Kaneohe). I knew that Hubert had been killed in a B-24 while flying a VP-11 mission out of England, but I did not know of the circumstances that

The Lockheed PV-1 Ventura patrol bomber was a derivative of the Lockheed 18 transport. Radar equipped, they were also flown by the USAAF in which they were called B-34s. (USAF)

Above: Honeymoon cottage, St. Augustine, 1944 (April 15-18).

Left: Bride and groom. Cutting the cake at the reception shortly after I blurted after the minister, "Now let us pray."

led to the crash. His parents came to NAS to pick me up and delivered me back to the BOQ after dinner and a delightful evening in their home. Except for the temporary additional duty flights, we were restricted in our travels and could not go beyond the radius of 50 miles from Jacksonville, so these three assignments temporarily relieved the monotony of repetitive instructional flights.

The original group of bachelors in the senior BOQ was more than 50 percent depleted by the time I started dating Grace Martin on 27 February 1944. I had a date with her every night until 14 April, the night before we were married, and the night of the bachelor party. My so-called friends in the BOQ brought me in at 5 a.m., and I just barely made the wedding across the river in Mandarin at Harriet Beecher Stowe's church at 4 p.m. on the 15th. We had not rehearsed the wedding, and I was so nervous while repeating after the minister, that I repeated after him when he said, "Now, let us pray." We are still married and are looking forward to our 39th anniversary in 1983. Corlett was my best man at the wedding, but the paths of our naval careers were about to separate, as he elected to begin training for duty in a carrier-based night fighter unit, while I decided to begin training for Pacific deployment in a B-24 or Privateer squadron (of four-engined landplanes). I was to begin flight training in this type of aircraft in October 1944 in Hutchinson, Kansas. We packed everything we owned, including wedding gifts of china and silver, into our 1937 Chevrolet coupe that we named "Blackie," and headed for our destination.

We got within 140 miles, and had travelled some very rough roads out of Olathe, when the car made a dreadful noise. I quickly stepped on the clutch pedal, turned off the ignition, and coasted to a stop on the shoulder. A kind man in a pickup saw I was in trouble and pushed us into Ottawa. All of the garages were busy, but I finally talked a mechanic into taking a look at it, even though he did not have time to fix it. My suspicions were confirmed when he advised us that two of the four bolts that held the engine in place had sheared off, and we would have to proceed by train. Before leaving Jacksonville, we had acquired a "dog-dog" (a cross between a terrier and a chihuahua) that we had named "Mitsie," and I had to crate the dog for shipment by rail and take it to the railway station. The train for Hutchinson had to be boarded at a junction five miles from town at 2 a.m. Sunday morning. We arrived just 24 hours before I was scheduled to check in for duty at the Naval Air Station. We had only a few changes of clothing, but we did have a place to stay as we had arranged to take over an apartment that had been vacated by friends who had finished training.

On Friday I received a telegram saying repairs had been made to my car. I advised my good friend Dennis Szabo, and he decided he needed to make a short field landing and take off at a field near Ottawa that was suitable for such an emergency landing. I got in the rear seat of the station SNJ, and off we flew. We landed in a wheat field at the edge of town, near the garage where "Blackie" was waiting to be picked up. I stepped out of the plane, watched "Denny" take off without incident,

The Consolidated B-24 in a scene from the U.S. Army training film, "Identification of U.S. Army Aircraft," 1942.

The Vought SB2U-1 scout bomber known as the Vindicator. (United Technologies — Vought)

then headed for the garage. The mechanic only charged me $25 for repairs, and I was back in Hutchinson for dinner that night with "Blackie" and all of our belongings. We remained in the apartment for only a week before moving into a two-story house with the Szabos. Denny's wife Rita (whom he had met in Australia) was pregnant, and my wife Grace was pregnant, and they were both on the same appointment schedule with the obstetrician, a Dr. West in Wichita. This required overnight outings at the Alice Hotel about every two weeks and provided many laughs. Although Grace and Rita entered the Catholic Hospital in Wichita at different times, their babies were delivered within three days of each other, and they were released on the same day. Our lot of living with two pregnant women was "a piece of cake," compared to what it was after the girls got the babies home. Denny and I were so busy sterilizing bottles that we could hardly find time to fly. All in all, training at Hutchinson was quite an experience, and it helped prepare us for almost any eventuality in a Privateer squadron.

From Kansas we drove "Blackie" back to Jacksonville where I received six weeks of operational training flying from the commercial airport. From there we drove to San Diego, where I received six weeks of advanced training at Camp Kearney. During this period most of the pilots in training who had their wives with them lived at Casa De Mañana in La Jolla, and most of us were already good friends before we arrived at the Casa. There were many memorable evenings during this six-week period, but it all came to an end for me in early July when I had to put Grace on a train and look for a buyer for "Blackie." My good friend Jack Poteat bought the car, drove it during his six-week training period, then sold it for what he had paid me.

On 4 July 1945 I made a simulated "transpac" flight. This was exactly like the real thing, except the pilot flew to the designated point of no return, then flew back to Camp Kearney. On this flight we took off at 9 p.m. and landed about 7 a.m. the following day. The night of our "transpac" was 10 July, and shortly after dark I went to the plane to give it one final check. There underneath the plane I found an enlisted man syphoning gasoline from the plane into five-gallon cans. I immediately called the guard on duty nearby and had him place the man under arrest. Three five-gallon cans were full and the fourth was about half filled. This thievery could have made a vital difference in flight. I

A USAAF B-24 Liberator bombing Nauru in the Pacific. The USN's aircraft were similar except for the front turret, as can be seen in the picture on the next page, taken on an unknown Pacific island. (USAF and Convair)

An early wartime Liberator cruising south past San Diego with the Convair plant and San Diego airport in the background and North Island Naval Air Station just out of the picture to the lower right. (Convair)

The PB4Y-2 Privateer was a single-fin variant of the B-24 with side blisters and forward-firing 20-mm cannon just below the insignia. (Convair)

did not know how much gas had been taken, or if anything else had been disturbed on the aircraft, so I got in touch with the plane captain. I told him what had happened and asked that he "top off" with gas, give the plane one final check, and have the crew ready for takeoff by 8:45 p.m. We were off on time at 9 p.m. and were instructed to climb to our assigned altitude of 9,000 feet and proceed to the first station ship 500 miles along our flight track. We encountered strong head winds at our altitude, and were 30 minutes behind schedule when we reached the first station ship, and 1 hour and 10 minutes behind when we spotted the second ship. When we reached the third ship, we were 1 hour and 30 minutes behind schedule, and 2 hours beyond the point of no return. At this point we received "transpacs" before, and I knew we could make it to Hilo, which was about 150 miles closer than Kaneohe. Fourth, I knew my flow-meter records were more accurate than the fuel consumption the plane captains were reporting by radio. At the sign of first light, I looked down and could see the white caps blowing right down our flight path at 15 knots or more. I let down to 1,000 feet and we coasted in to land at Kaneohe with 18 minutes fuel remaining, after just over 15 hours in the air. It would have been much closer, if we had been short 20 or 30 gallons of gas. The Wing Commander wanted to know why we did not follow orders and return to San Diego. I said, "What orders?"

My crew was about to finish pre-deployment training at Kaneohe when I received orders to report to San

The first section of the 18-plane Privateer squadron, VPB-108 led by LCDR Joe Hill, CO, in the lead plane, near Okinawa, December 1945.

a message from San Diego control to return to base. The radioman brought the message to the cockpit, and I told him not to acknowledge that we had received it. Although none of my crew had ever made a "transpac," and they were scared half to death when they learned that I was ignoring the message and proceeding on to Hawaii, I had several reasons for making the decision. First, I knew that if we returned to base there would be morning fog in San Diego and we would not have enough gas to fly over the mountains to El Centro and land. Second, I knew that 98 percent of the time the trade winds were blowing 15 to 20 knots for the last 1,000 miles toward Hawaii. Third, I had made two Diego to appear as the key witness in the court-martial of the poor sailor who was apprehended stealing gasoline from our plane. On 29 July I was flown from Honolulu to San Francisco, then to San Diego the next day. I appeared at the trial on 2 August, flew to San Francisco on the 4th and back to Honolulu on the 6th. Before we deployed on the 11th, B-29s dropped the A-bombs on Hiroshima and Nagasaki, and we knew the war against the Japanese was rapidly drawing to a close.

On 11 August, we flew to Johnston Island, where we joined several other Privateer crews, including Hugh "Buck" Rogers, in a farewell party before leaving for

Kwajalein the next day, then on to Tinian on the 13th. On 14 August, I flew from Tinian to Guam to meet our Wing Commander and inform him that I was relieving LCdr. Robert "Buzz" Lefever that day at Iwo Jima. The change of command was without ceremony, as I told Buzz, "I accept VPB-108 as it is, and you are officially relieved." This was 15 August, the day the Japanese threw in the towel. What I had assumed was command of an 18-plane squadron of Privateers, with 20 mm cannon in the nose. Six of the planes were physically located at Iwo Jima, six at Peleliu in the Palau Islands, and six at Tinian, our home base. Although the unofficial surrender had been made, the Japanese in the outer islands did not know this, and our patrol planes were being fired upon until early in September, when the official surrender of all Pacific forces was made at Truk Island on the 2nd. The formal surrender of all forces, unconditionally, came aboard the USS *Missouri* on 15 September 1945, and thus ended almost four years of brutal warfare between the United States and Japan.

Although most of my stories and those of my contemporaries are about war in the Pacific, there were many cadet-trained pilots fighting in England and the European theater. In fact, many of our old VP-11 pilots were deployed to England in B-24s for their second deployment. They included Poulos, Muckenthaler, Willis, Reese, and Enloe. Hubert Reese was one that did not make it back. He was either shot down, or crashed at sea, but nothing was ever recovered of his plane or his crew.

There were many other cadet-trained pilots who flew offensive missions to the Continent from England. One of these involved a special task group under Cdr. Jimmy Smith; a Hugh P. "Rosy" Lyon was a pilot on the mission. Special Attack Unit One of this group flew to Dunkeswell Naval Station in England on 24 July 1944, and two days later to Fersfield, in East Anglia, which was to be the operations base. The attack unit consisted of Lockheed PV "Mother planes," and a B-24 (PB4Y-1) drone plane. The mission was designed to prove that drones filled with torpex (an explosive far more powerful than TNT) could be flown in by remote control with "Mother planes" to wipe out the buzz bomb sites called "Spam Factories." (I guess our pilots

PBY-1 with North Island in the background, 1940. (Convair)

hated them as much as they hated Spam.) The primary target of the first mission was a "Spam Factory" on the French coast and involved the number one son of the U.S. Ambassador to England, Joe Kennedy, Jr. He and Bud Willey, two pilots in the attack unit, had volunteered to take off in the torpex-laden drone and bail-out through the nose-wheel well once all systems were checked and the drone was under positive control of the "Mother plane." On the afternoon of 12 August 1944 the drone and two "Mother planes" took off and rendezvoused at 2,000 feet altitude to check out all systems before Kennedy and Willey bailed out. Rosy Lyon was flying the "Mother" PV just behind and slightly above the drone, which was in the middle of the checkout, when he heard a gasp from his co-pilot, Buser, as the exploding drone filled the sky with flame and debris. There were two distinct shocks as Rosy's plane was thrown on its back, but he had enough altitude to recover from his inverted position. And he and the other PV flew back to Fersfield and landed. The total time of the flight according to Rosy's log, was .7 hours. By the time the attack unit got another PB4Y drone rigged for a mission, General Eisenhower's forces in France had overrun the "Spam Factories." This particular attack unit made one drone attack with a PB4Y when Lt. Ralph Spaulding from one of the Fleet Air Wing Seven squadrons at Dunkeswell took off solo, climbed to 2,000 feet, turned the control over to Lieutenant Wherry in a "Mother plane," and bailed out through the nose-wheel well. Wherry controlled the drone until it was a few miles from target (a German submarine pen), and then turned control over to Rosy's plane, which was to home in the drone to the target by a television camera system. Before the drone reached the target, the television went blank (probably shot out), so Rosy put the drone into a dive, where it detonated into a fuel dump, which was not even close to the target. Shortly after this the Navy's drone program in Europe was discontinued, and this particular attack unit was returned to duty in the U.S.

PART III
RECOLLECTIONS

Chapter Six
Prisoners Of War

When I think back about the Japanese attack on military installations and ships in Hawaii, I have a certain amount of respect for naval officers and pilots who planned and executed the attack. In fact, if the planners could have had the foresight to include a few troopships loaded with invasion troops, most of the U.S. forces on the island would have become prisoners of war, and the war would have lasted much longer than it did. Although I have learned to forgive the pilots who made the attack, and respect them for what they did, I can never forgive the Japanese as a nation for allowing the cruel, inhuman, and atrocious treatment of our U.S. prisoners of war.

There may have been many reasons why such treatment was condoned, but there was absolutely no excuse for it. The Oriental mind is not easy to compare with the mind and thought processes of people in the Western world. Their life styles are different, and their wants and needs are different. Their morals, education, and religions cannot be compared. Starvation rations for an American would be acceptable fare for an Oriental, and there are so many of them living below what we consider the subsistence level, that they have little regard for human life, or the rights of others. This was particularly true with regard to the Japanese feelings for the Chinese, and was clearly demonstrated when they massacred 42,000 Chinese civilians in Nanking at the beginning of their invasion in 1937. The Chinese learned quickly that the best policy for them was to hide their valuables such as silks, linens, gold, silver, and jade, and make orderly retreats without resistance. To pursue any other course would mean certain death, because the Japanese were not going to spend any of their war effort feeding Chinese prisoners. Another reason for the Japanese attitude toward U.S. prisoners was that they had no respect for criminals in their own country, and they considered prisoners of war lower in the social order than criminals. Furthermore, all Japanese military men had been thoroughly indoctrinated in Bushido, a feudal-military code of chivalry which valued honor above life. It was a distinct honor to die fighting for the Emporer and their country, and it was better to commit hari-kari than to be captured as a prisoner of war. This was particularly true among Japanese officers who looked disdainfully upon all U.S. prisoners, and delegated administration and supervision of them to subordinates who were rude, crude, overbearing, and sadistic.

The first Americans to become prisoners of war were the Marines assigned as the Embassy Guard in Tientsin and Peking, China, on 8 December 1941, shortly after the Pearl Harbor attack. At about the same time the U.S. Gunboat *Wake* was captured at the dock in Shanghai, and Lt. Commander Smith and his crew became prisoners. The next group of prisoners were the U.S. forces on Guam, which surrendered on 10 December. None of the natives on Guam were taken prisoner, but they were forced to perform all kind of unpleasant duties for officers who built some fine frame quarters and an airstrip on land that is now NAS Agana, Guam. The quarters were built without windows or doors, but were completely enclosed with fine mesh screening to allow the ocean breezes to blow through them and keep them cool. They also had canvas flaps that could be lowered to keep the rain out, and had screen doors on three sides. These quarters may still be used today, and, if so, stand as a tribute to the fine materials that were part of their construction. The natives of Guam will always hate the Japanese who controlled the island during the war years and caused them so much suffering and deprivation.

The next large group of people taken into custody after the war began were Japanese, first on the island of Oahu, and then in the West Coast states of the U.S. Altogether, more than 100,000 were rounded up and interned for the duration of the war. Many of them had been under surveillance by various intelligence groups long before the war started, and they were taken into custody to protect us from spying and sabotage, and possible assistance in invasion attempts. However, their internment served to protect them from hostile attacks by U.S. servicemen and civilians who had lost friends, relatives, and buddies during the 7 December attack. They were never treated like prisoners, as they were allowed to take as much of their personal belongings as they could carry, and they were always provided with good clean beds and plenty of good wholesome food. They were allowed to get as much exercise as they wished, there were no communications restrictions, and they received medical attention.

When Shanghai fell to the Japanese in the early days of the war, the people in the International Settlement were allowed to remain there under the supervision of the Japanese guards and they lived about like they did before they became prisoners. If they had money, they were allowed to make purchases of food and tobacco by bribing the guards.

The next large group of Americans to become prisoners of war were the military and civilian forces which surrendered on Wake Island on 23 December 1941. They remained prisoners on the island until they were herded aboard the *Nitta Maru* on 12 January 1942 for the trip to Yokohama. However, not all of the prisoners got aboard the ship, as more than 200 of the civilians were forced to stay on the island as slave laborers in violation of the Geneva Convention. There is no official record that these men ever existed, because they were executed and buried before our forces recaptured Wake. The living conditions aboard the ship were deplorable, as the men were dumped in the ship's holds without heat or plumbing. They were fed small amounts of rice and water twice a day, and all elimination was done in buckets. Rules were posted in each hold, and violation of any rule was punishable by death, so the prisoners were afraid to do anything. There is no record of how many safely made the trip to prison camp, but the war crimes trials established the fact that at least five enlisted personnel were beheaded in one gruesome ceremony during the trip. About two percent of the prisoners were separated from the main group in Yokohama and sent to Shikoku, but all of the others were sent to an old Chinese army camp at Woosung, China. It was not far from the junction of the Whangpoo and Yangtze Rivers and was surrounded by a high electrically-charged fence. Before the Wake Island prisoners left the island, at least one of them, cadet-trained Marine fighter pilot John F. Kinney, began making plans for an escape.

As our Army and Naval bases fell in the Philippines, many entire families became prisoners, and they were stripped of all of their valuables including watches, jewelry, money, and everything in their safety deposit boxes. Most of these people became prisoners in makeshift camps or buildings at the bases where they were captured. However, the greatest number of Filipinos and Americans to become prisoners of war up to that time was when more than 75,000 surrendered on 9 April 1942, at Bataan. Of the 70,000 who were able to begin the 65-mile "death march" to Camp O'Donnell, more than 10 percent died. The only cadet-trained pilot I know who successfully made the march is Andy Reid (Class 128 C), who now lives in Pensacola, Florida. To this day he admits that he does not know how he made it. Many of the survivors owe their lives to their stronger comrades who carried them in improvised litters when they could go no farther. Even so, more than 400 marchers died every day. Less than one month later, on 6 May, General Wainwright surrendered his garrison at Corregidor. They were taken to an unfinished Philippine army camp at Cabanatuan, where more than 15 percent died during the first two months. In both of these large groups, survivors were forced at gun point to dig graves and bury their buddies who had died along the way. The cruel, inhuman, atrocious treatment of prisoners in the Philippines should remain with the Japanese people and their conscience forever. War was no excuse for this beastly behavior, and atrocities by guilty-conscience soldiers are still coming to light as this is being written. Some of these were so gross that they were never mentioned at any war-crimes trials, or in any history books. One of these soldiers with a guilty-conscience recently admitted that he was one of several truck drivers who were ordered by Lt. General Shiro Ishii to run over and kill 40 prisoners who were trying to escape an experiment with bubonic plague that was being forced upon them.

On 10 September 1942, one of our best and most experienced pilots in VP-11, Lt. Carl Clark, was flying out of Ndeni in the Santa Cruz Islands when he encountered a large Japanese troopship at the end of his 700-mile flight sector. As narrated earlier, the ship was alone, and Clark decided to make a bombing run on it. Little did he know that the ship had catapult-launched eight float Zero fighter planes about 15 minutes before he began the run. Before he pushed over, he was attacked by four Zeroes, who shot both engines out and wounded one of his crew members before he could make a deadstick landing. The Zeroes continued to strafe them in their airplane until they ran out of ammunition. All crew members got out of the plane safely. Soon a destroyer picked them up and rushed them to a heavy cruiser which was the flagship of the Task Force that made up the 4th Fleet. It was near sunset when the destroyer put Clark's crew in a whale boat and rowed them to a "Jacob's ladder" which led to the after deck of the cruiser.

When they got aboard, Clark said he had never seen so many Japanese on deck, and he thought they had been called to some kind of parade to witness the execution of the entire crew. He found out later that the reason for the all-hands formation at sunset was so that the crew could face Tokyo and pay homage to the Emporer and Japan. Each of Clark's crew members was brought before a podium on the after deck and questioned at great length, but none of them gave answers other than name, rank, and serial number. Clark was placed in a 3- x 5-foot cell under a 5-inch gun mount, and he was not called for interrogation until midnight. He still had the idea that they were all to be executed, and he lost about five pounds in sweat before six guards came after him. He guessed the reason for the large number of guards was to prevent him from jumping over the side and committing suicide. The interrogation that night was Clark's introduction to the type of punishment he would have to endure for three years. For the 1½ years, Clark and his crew were imprisoned by the Japanese Navy at Ofuna, Japan. Initially, they were placed in solitary confinement, without adequate clothing, in unheated spaces, and they were not officially listed and registered as prisoners of war, so that friends and relatives did not even know if they were alive. (Clark's family did not know he was alive until Christmas 1944.) Clark was interrogated and beaten every day for the first six months, and he was given a small amount of rice and water twice a day. He dropped from 185 lbs to 113 lbs during that period, and had cuts and bruises all over his body. He also lost many of his teeth and some of his hair. The food was hardly enough to keep him alive, and the exposure to the cold at night was almost unbearable. However, early in December he got more clothing which the guards said came from Wake Island prisoners. Although the clothes were mostly rags, he said he could not have made it through the winter without them. Throughout all of the beatings and interrogations, Clark would tell them nothing but his name, rank, and serial number. During the second six-month period, the beatings became less frequent, but came at all hours of the day or night. Prisoners were not allowed to talk with each other during the brief exercise periods in the yard, and if they were caught whispering they were given exhibition beatings in front of all prisoners. During his last six months at Ofuna, interrogations dropped to once a week, and there were no more beatings in private. The long hours of solitude were hard to take, but during that last six months at Ofuna, Clark began to think that he was going to survive the cruel punishment and make it through the war. Late in December 1943, Clark and three other prisoners at Ofuna were taken to the prison in Omare, Japan, which was to be their home until 10 September 1945.

As the big gates at Omare opened to let the four men in, they were greeted by the sight of a prisoner hanging from a small tree with his toes barely touching the sloping ground. He had been sentenced to ten days in this position from sunup to sundown, because he had been caught stealing while assigned to a work detail. This was not a welcome sight for the new arrivals, and they wondered if they were going from bad to worse conditions as prisoners. Clark learned that there were about 600 at Omare, including British, Australian, Canadian, and U.S. prisoners, but they were registered as official prisoners so that relatives were notified that they were still alive. The routine at Omare was a great improvement over that at Ofuna. Prisoners could mix freely with others in the yard during exercise periods, and there were no restrictions about talking. Life became much more interesting as prisoners learned all about each other, and they got late news about the war as new ones came in. They had more frequent and longer lasting exercise periods. The physical living conditions were better and more comfortable than they had been at Ofuna, and they received enough food to stop their weight loss. Although conditions at Omare were much improved, Clark soon learned that the disciplinary head for his group was an arrogant Sergeant by the name of Wanatabe. He charged all rule infractions and disciplinary problems to the senior man, Commander Mayer (senior survivor of the cruiser *Houston*). He loved to beat on senior officers. Carl was unable to learn if Wanatabe was ever punished by the War Crimes Commission, or if he was ever charged with any crimes, but he knew was that he should have been punished for several. For the prisoners at Omare and other prisons on the main islands of Japan, there was no hope of escape. There were no friends outside the prison walls, and there was no place to go or hide. So the only thing these prisoners had to look forward to was the end of the war and their release.

Not long after Clark was moved to Omare, he was joined by a fellow PBY pilot, John Marshall "Buck" Arbuckle, a Pensacola classmate of mine in 130 C. Like Clark and myself, he had been sent from Pensacola to a PBY patrol squadron for his very first duty station, and like Clark he had been shot down by Zeroes while on long-range patrol. However, Buck was the only survivor of his 11-man crew. He was shot down about 14 months after Carl, and remained a prisoner for about 22 months. Buck brought much late news about the war in the Pacific. The battle of Guadalcanal had been won by the Americans, and Henderson Field was one of our most active land bases. However, we lost many ships, planes, and pilots, including the carriers *Wasp* and *Hornet* in winning the battle. U.S. forces were moving up the island chain and were concentrating on knocking the enemy out of his major staging base at the northern tip of Bouganville. Many of the islands lost in the early days of the war had been recaptured, and many more new carriers with full complements of pilots and planes were reporting to the Pacific. Even some new battleships and cruisers had joined our forces to form large carrier task forces that could control the air space over any given area for long periods of time and allow our amphibious forces to make successful landings. This would eventually mean recapture of Guam, the Marianas, and the Philippines. Buck also brought the news that Admiral Yamamoto had been shot down and killed while flying to an inspection tour of Rabaul. The arrival time at the island to be inspected was deciphered from an encoded message, and a squadron of P-38 Lightnings from Henderson was assigned the task of intercepting the Admiral's flight. The P-38s were under the command of the Army's Major Mitchell, and the task was not expected to be an easy one, as he expected Yamamoto's plane would be protected by at least a full squadron of Zeroes. As it turned out, only six planes were escorting him, and the P-38s accomplished their mission with the loss of a single plane. Buck's arrival at Omare made life much more interesting for Carl, because their training periods at Pensacola overlapped and they had a lot in common as Patrol Plane Commanders. They became great friends at Omare and remain so to this day, although Carl lives in Pensacola and Buck lives in Houston.

Buck had not been at Omare but a few weeks when he was given a cadet-trained cellmate. He was Marine fighter pilot Gregory "Pappy" Boyington, who was one of the first-year aviation cadets recruited from the Seattle, Washington, area in 1935. He served out his contract and was a Marine flight instructor at Pensacola in July 1941 when he resigned his commission to join the American Volunteer Group to fight the Japanese in China. He remained with the Group until they disbanded about a year later and had credit for six Japanese planes. Because Clark was one of the pilots in VP-11, along with Willie Howe and myself, who wanted to join the AVGs, he and Boyington had a lot of catching up to do about the Flying Tigers. Greg brought even later news about the progress of the war, and indicated that our forces probably would be operating from the Marianas before long. Once bases were established there, our long-range B-29s could be bombing the industrial cities of Japan, and the prisoners at Omare later were to learn this is exactly what happened. Although Greg shot down only six Japanese as a Flying Tiger, he shot down five on his first mission as a Black Sheep, and had a total of 28 planes to his credit when he was shot down. This total was to be tied only by fellow Marine fighter pilot Joe Foss, who was a Pensacola cadet in Class 150 C. Although Carl and Buck were glad to hear about the progress of the war from Pappy, they soon learned what they had suspected all along, that Marine fighter pilots and Patrol Plane Commanders did not have much in common. They were fighting the same war, and they happened to be in the same Japanese prison, and they realized that they had to do the best they could to cooperate and get along with each other, if they were to get out alive. Pappy was about the only person at Omare who realized what a big "wheel" he was, and this did not make life any easier for the PBY pilots, particularly for Buck, his cellmate.

Although none of the prisoners had any idea how long the war would last, as new prisoners came in and told their stories about the war's progress, they began to get the feeling that the end was in sight. This made them more determined than ever to stay alive and survive their ordeal so that they could get out of prison.

Whereas there was absolutely no hope of escape for the prisoners on Japanese islands, the prisoners in China were always thinking about it, because they knew if they could get far enough away from prison camp they would find friendly Chinese who would help them. The men at Woosung who came from Wake Island had not been in prison long before they planned to attempt an escape, particularly the senior officers who were not under constant surveillance. The first attempt included Commander Cunningham, Wake Island Commander; Dan Teeters, senior contractor's representative; Lt. Commander Smith, Gunboat *Wake's* Commander; and Commander Woolley of the British Embassy in Shanghai. The attempt had to be made under cover of darkness, and escape had to be made digging a hole under the electrically-charged fence. Dan Teeters was elected to produce the shovel and do the digging, and on the appointed night after the last bed check, he dug a hole large enough and deep enough to allow four adults and a Chinese boy to get under the fence without being electrocuted. The Chinese boy was supposed to help them find some friendly Chinese and act as their interpreter. They had some general idea where Woosung was located, and they knew they would have to get out in the farm country well away from the prison before they could find people who would risk helping them. Chinese near the prison knew that if they were caught helping prisoners escape, it meant instant death for them. It was after midnight when they got clear of the fence, and they could not get far enough away from Woosung by daylight to get help. When they did find a farmer who said he would help them, he locked them in his barn and reported to prison authorities that he had captured them. Their freedom had not lasted one half of a day, and they were back in Woosung charged with the crime of escape. They all expected to be executed, but were subsequently brought to trial and sentenced to ten years in prison. The sentence was less severe than they had anticipated, and they were incarcerated in Shanghai prison where conditions were better than they had been at Woosung. Other attempts to escape were made by small groups at Woosung, but none of them were successful until Lt. John F. Kinney and three of his buddies escaped while being moved from there to another prison in May 1945.

Kinney had been planning and thinking about escape since he had become a prisoner on Wake. He was always on the alert for scrounging materials and was saving them for the opportune time when escape would be possible. Shortly after they became prisoners at Woosung, the Wake island officers, including Kinney, were furnished with small radio receivers so that they could hear the Japanese propaganda about their successes in the Pacific. There were three receivers in all, but the only station they could get was the propaganda station operated by the French in Shanghai. Kinney soon modified one of these sets into a short-wave set that could receive broadcasts from San Francisco, and he kept his fellow prisoners advised about the real progress of the war. When he advised them about our success in turning back the invasion forces at Midway, Japanese guards who could speak English got wind of these reports and confiscated all radio equipment, so Kinney had to start thinking about what he needed to build another set.

As a boy he had taken the receiver apart many times on the hand-cranked battery-powered telephone and put it back together, so he knew what he needed to begin building an earphone. One of the things was a magnet, and he remembered that all speedometers had to have magnets. He prevailed upon a POW by the name of Fuller, who worked in a truck repair shop, to get him a magnet from a speedometer. Fuller was also able to get a damaged ignition coil which had more than enough fine wires required for the earphone. The other parts needed were easy, as he used a Nescafé powdered coffee can he had received in a Red Cross package as the base for his earphone, and it also provided a means for concealing it as it looked like any other coffee can. He made electromagnets out of two nails, which he wound with about 5,000 turns of fine wire, and soldered them to the magnet which was mounted on a piece of wood that fit the base of the can. He used lead scrapings and the tin from toothpaste tubes to make his solder. He used the seal from an English tobacco can as a diaphragm at the top of the electromagnets which actuated it with a weak pulse of current. Kinney did all of his work in secret, because he did not want anyone to know about it in case it failed, or in case news about the project leaked out again to Japanese guards and caused another shakedown. Also, he did not want to subject his roommates Lt. Bob Greeley and Ens. Joe "Turkey" Davis to knowledge that could get them in trouble. He knew the Japanese had methods that would make anyone talk, if they persisted long enough. The work was time consuming, and because it was all done in privacy, it was done while his roommates were asleep, or carried out in the benjo (latrine) which had separate stalls. It took over six months for John to get 5,000 turns of wire around the nails, and after he completed the earphone, he had no way of testing it, so he started building a crystal set. He had high hopes of making a small crystal set work, because he remembered reading about a freakish case where a man was able to receive broadcasts due to the peculiar arrangement of the fillings in his teeth. He assembled the crystal set in a 4- x 5-inch powdered milk can that also came in a Red Cross package. Kinney had quite a time getting his crystal set together, and at first nothing seemed to work, but he refused to give up. He established his private radio shack in a benjo stall, with an antenna on the roof, and a stake wrapped with wire in the ground. These terminated at a nail in the special benjo stall where he could hook on to the two special nail ground/antenna terminals. Because his private radio shack was not for his use exclusively, he sometimes had to wait to get into the stall. When he could not get his crystal set to work, he turned to his good friend, Capt. Herbert C. Freuler, who was a chemical engineer.

With Freuler's help and some new material, he reassembled the crystal set and began picking up some faint sounds. This was encouraging, but he suspected that his earphone needed more amplification, so he re-spaced the poles and put an additional 11,000 turns of wire around each nail. The next time he tried the set, he got very good reception, and he began to get the urge to visit the benjo at the same time as the Vichy and Russian English news broadcast periods. He was able to get news of the war in the Pacific and in Europe, and one of the first news items he picked up was a Russian broadcast that the Italians had surrendered in September 1943. From then on during the rest of his captivity he kept a privileged few of his fellow prisoners advised about the progress of the war. His radio helped him and three of his buddies plan and execute their successful escape. On 10 May 1945, when many of the Woosung prisoners were being moved by train to a prison camp further north, Kinney and three of his buddies managed to elude their guards and jump out of the window of the fast-moving train into the darkness. After many close calls with a search party that was looking for them, they reached United States Armed Forces at Kunming. They had travelled through enemy territory by foot, horse, and boat for a distance of 700 miles. Kinney's group of escapees arrived in Washington, D.C., on 10 July 1945, and reported for duty at Headquarters USMC. They remained on duty in Washington briefing Army, Navy, and other intelligence activities until the 23rd, and then on the 25th, orders were modified for duty at MCAS (Marine Corps Air Station), El Toro, California, with 60 days delay in reporting.

Although the prisoners at Omare did not have a radio to keep them advised about the progress of the war, they did get news when French prisoners arrived. In the spring of 1945, they began to see some B-29 bombers on the way to their targets, and they guessed that they came from Guam and the Marianas. About two months before the war was over, several U.S. carrier planes came in low over the prison yard and dropped packages of food and clothing. Some of the packages missed the yard, but most of them found their target and they were a joyous sight. All of the prisoners, including Clark, Arbuckle, and Boyington, knew the war was almost over. On 10 September 1945 they were released and were headed back to the good old USA.

This chapter has covered only people who were made prisoners by the Japanese in the Pacific because this is where almost all of the cadet-trained pilots became prisoners, and it has been from some of them that I received first-hand accounts about how they were treated.

Chapter Seven

Hair-Raising Incidents In The Cockpit

This chapter must begin with a general discussion about pilots, aircraft, and what constitutes a hair-raising incident. From the beginning of aviation, aeronautical engineers, aircraft manufacturers, and pilots have been working on theories, production of equipment, and development of procedures that would make flying safe. Almost every fatal accident has caused some change in equipment or pilot procedures, and this is a continuing process. All new aircraft are designed and built with the benefit of accumulated knowledge and experience of millions of hours of flight. Many test pilots have given their lives toward the building and production of better airplanes and the refinement of air transportation. During my lifetime we have progressed from World War I "Jenny's" to long periods of flight in space in the *Columbia*, with the manufacturers of aircraft and space equipment now working on their drawing boards to develop products that will be vastly superior to those we are using. Although the airplane was thought to be a very dangerous machine in the early days of flight, it is now the safest mode of transportation per mile traveled. When one considers the incredible amount of aircraft traffic around our busiest airports such as O'Hare, Los Angeles, and Atlanta, the air transportation safety record is inconceivable. But even with such a great safety record, there is hardly a minute of the day that passes when there is not a hair-raising incident in some cockpit. Some near misses are reported, but most of them are not, and most pilots never know they are close to a collision, so the hair does not rise.

With the previous paragraph as background, we need to get a clear and concise definition of what constitutes a hair-raising incident. This is a situation that is created or happening in the cockpit of an airplane when the pilot has some doubt if he is going to get himself, his crew, his passengers, and his equipment to a position where he can make a safe landing. The pilot may be in a hair-raising situation, but, if he does not realize this, it does not become an incident.

The hair-raising incidents in the cockpits that I am most familiar with are those that happened to me as a pilot in 20 years and almost 6,000 hours of flying while I was in the Navy. I will record only those incidents where the hair was standing. There may have been situations created while I was flying commercially that could have caused the hair to rise, but I was not aware of them, and they did not become incidents.

The first incident that happened to me in an airplane, when I did not think I would get back on the ground alive, occurred at Pensacola, on my last instructional flight with Ensign Monk Russell, although he was not in the plane at the time, as I made six shots to the 100-foot circle, as already related.

The next incident happened less than a month after I reported to Patrol Squadron 23 at Ford Island. On 24 September 1940 Ensign Eddie Harshutz and I took off from the Dry-dock Channel at Pearl Harbor for a four-hour instrument familiarization flight. Eddie was a first pilot, qualified to take a plane out on a limited number of missions. I was his co-pilot while he was practicing instrument flying under the hood. The hood was a piece of canvas shaped to fit the pilot side of the cockpit, and it was fastened with metal fasteners so that the pilot could not see out of the plane. As the co-pilot, I was supposed to be the look-out and make sure the areas we were flying in were clear of aircraft and all other obstacles. When we were practicing instrument flying, we were usually flying at 5,000 feet or more, above the clouds and the mountains of Oahu, and this day was no exception. As co-pilot and look-out I would make sure that the area ahead was clear and put the plane in all types of unusual positions so that Eddie could make a recovery while on instruments. We were about three hours into the flight when I put Eddie into a graveyard spiral. In this type of maneuver, the plane starts out in a turn (left or right) and a shallow dive. As the pilot on instruments sees the airspeed building up, the normal tendency is to pull back on the yoke to decrease the speed, but this only tightens the turn and allows the airspeed to keep building up. In this particular case, we were in a diving left turn at 5,000 feet, and the port (left) engine quit cold. This aggravated the turning moment of the plane so much to the left that the plane was almost in a spin. I motioned for Harshutz to pop open the hood, and I took over control of the plane, throttling back on the good engine to try to get out of the turn. Once the hood was down Eddie took control of the plane and we both began looking for a place to land, because we both knew we could not maintain flight for very long on one engine. The engines on the PBY-1s could not be feathered and the windmilling prop created such a drag and so much vibration that the plane could not fly on one engine for long. Fortunately when Eddie

PB2Y-2. (Convair)

got the plane under control, we were less than 15 miles from the protected waters of Kaneohe Bay, so we proceeded to land close to shore and anchor. We broke out a rubber life raft, rowed ashore, and found a telephone to call the squadron at Pearl Harbor for help. The engineering officer sent two chief mechanics with a complete set of spark plugs, along with some other spare parts and tools, and they had the engine running by 4 p.m. We took off from Kaneohe and were back in the Dry-dock Channel by 4:30.

The next incident in my cockpit, when things got a little hairy, happened during the period when we were flying PBY-5A amphibians out of Barking Sands on the island of Kauai in preparation for the Battle of Midway. The landing strip was named Barking Sands because the land on that side of the island was so hot and dry that the sand made a barking sound when you stepped on it. George Enloe was my co-pilot and navigator because Ken Kemper, my regular co-pilot and one of the best navigators in the squadron, had some ailment and was grounded. Flying from Kauai on a land-based strip extended our patrol range toward Midway so that we were only 250 miles from Midway at the end of our sector. On 22 May 1942, we were returning from a 12-hour patrol. In preparation for the landing, we had both failed to read off the landing check-off list, and this is where the trouble started. It was a good smooth landing, but the instant we touched down I knew we had failed to release the parking brakes. In the PBY-5A, if you put much pressure on the top bar of the rudder pedal, you locked the parking brake on the side where you had put the pressure. Thank God the brakes were locked on both sides, or we could have taken out a row of airplanes that were parked along the runway. As it was, we blew a tire on the right side as we skidded down the gravel strip. There was absolutely no excuse for this incident, because if we had followed the required procedure of reading the check-list, it could never have happened. One of the items on the list after "gear down and locked," is "release parking brakes."

There was one other incident, which was caused by bad weather, and it happened while I was patrolling around the big island looking for Japanese who could have been put ashore by submarines. I was one of two planes assigned to the converted four-stacker destroyer *McFarland* (DD237/AVD14), which was being used as a seaplane-tender in Hilo Bay. The routine was to fly around the island clockwise beginning at dawn, then reverse the patrol in the afternoon landing about dark. Each circuit of the island required about four hours, so pilots and crews were flying about eight hours every other day. On 11 November 1942, I had completed the clockwise flight and landed about 1000. We had lunch we would be most fortunate to locate the tender under such conditions, and the high winds would make a landing hazardous, so I turned the plane around and headed for the Kona (lee) side of the island. My Chief Mech was a full-blooded Hawaiian who used to live on the big island, and he knew a lot about conditions all around there. I discussed with Lupenui the idea of making an open-sea landing along the Kona coast, and we both decided this was much better than any other alternative we had. He cautioned me about landing close to shore, as there were large ship mooring buoys

PBY Catalina. (Convair)

aboard the tender and prepared for the afternoon flight. We reached the plane about 1400 and took off for the counter-clockwise circuit. We did not take on additional fuel for the second flight each day. We had completed about three hours of the afternoon flight when we ran into a severe storm. The wind was about 50 knots and visibility was less than ¼ mile. We tried to raise the tender by radio to get a weather report at Hilo, but there was so much static and atmospheric disturbance that we could not make contact. I decided that anchored less than a mile from shore. We reached the area about 2000 (two hours after dark), and I began to let down for a landing at least two miles from shore. It was a very dark night, and the island was blacked out, but I eased the plane down at less than 200 feet per minute with power on until I could feel the air cushion as we neared the water (the PBY wing span was so great that the pilot could always feel this cushioning effect as the plane reached a point three or four feet from the water). When I felt the cushioning effect, I chopped the throttles, and we touched the water the first time. The landing was smooth enough, but we made a series of bounces as we came into the huge swells. We taxiied toward shore until we found one of the large ship mooring buoys and tied up for the night. Lupenui said he knew of a place along the coast where we could make a call to the Hilo Harbor Master to notify the *McFarland* that we were down and moored safely for the night and would return in the morning. I set up machine-gun watches for the bow and the two waist guns in case a submarine surfaced near us during the night, and Lupenui and I went ashore in a life raft. We found the place he had in mind, made the call, then returned to the plane. The night was uneventful, and we began our takeoff shortly after daylight. However, the swells were so bad that we damaged a wing-tip float before we were safely airborne. We flew back to Hilo Bay and advised the tender that we were proceeding to Kaneohe, because of the damaged wing-tip float, but we would request a plane to relieve us in Hilo that same day. We landed at Kaneohe and made it safely to the recovery ramp by putting a man on the wing with the undamaged float.

The next incident happened in the South Pacific during the Solomons campaign. It was 10 September 1942, the day Willis and his crew were picked up at sea, except for his plane captain (Frankovich), who was wounded when they were shot down, died the second day, and had to be buried at sea. It was also the same day that Clark and his crew were shot down by Japanese Zeroes on floats and were taken prisoners. On this date I had the patrol sector which crossed a sector covered by Japanese patrol planes, and contacts were made almost every day. I decided to run my patrol at 5,000 feet, so that I would have an altitude and speed advantage if I saw the enemy plane. We were on the return leg about 300 miles from Ndeni when we sighted the plane about 15 miles away flying at about 1,000 feet. I maintained my altitude until I got closer, and then I began my dive toward him. By the time we got within a few miles of him, it was obvious that he had spotted us, because he began to pick up speed in a shallow dive. We did have a good speed advantage and were able to overtake his plane and give the bow gunner and port waist gunner several good bursts, knocking out one of his gun positions. However, we did no damage to his engines or vital controls, and he began to pull away from us after we made our initial run. From the first sighting until he pulled away from us, the encounter had lasted less than 15 minutes, but it had taken us out of our patrol sector. We climbed to 1,000 feet and headed for Ndeni. The only problem was that we were not where we thought we were. The upper winds at 5,000 feet most of the day had taken us far out of our patrol sector, and the run on the Japanese had only aggravated the situation. We missed the island of Ndeni completely, and when we discovered where we were, we had to fly directly into a 35-knot wind for more than two hours before we could see the island. By then it was about dark when we landed in Graciosa Bay in a strong crosswind. We had been out almost fifteen hours and we were about out of gas when we reached our mooring buoy. With Ken Kemper as my co-pilot during the first year of the war, this was the only time we were ever lost, but on 10 September we were lost for most of the day!

This episode taught me several lessons that should have been obvious all along. First, the best pilot in the world is of little value if he is lost. Second, pilots should restrict the missions for their planes to those for which they were designed (you can't make a fighter plane out of a PBY). After Clark's plane was shot down, Admiral Fitch put out the word that patrol planes would be limited to long-range patrols and anti-submarine work for which they were designed, unless they were sent on special missions with bombs and torpedoes.

The next incidents all happened while I was stationed at the Naval Air Station in Jacksonville, Florida, as a flight instructor attached to the Advanced Training Command. They all happened at night, while I was teaching young Ensigns who had just received their wings how to land and take off in the St. Johns River. Each instructor had three Ensigns aboard, and we gave each of them one hour of bounce drill. Before each night's flying, one instructor was assigned as boat officer for the night. About an hour before dark each evening, the boat officer and a coxswain would take a motor launch out in the river and position themselves so that night flying instructors could land and take off somewhere near a string of lights that would be trailing behind the anchored motor launch. This system worked fine as long as there was a good steady wind, but if it was swirling and slight, or changing, conditions were ripe for an incident. The boat officer could turn off the green trailing lights, which meant for pilots to land at their own risk. In this case, the only lights visible were the running lights and the stern light on the boat. The boat officer could also turn on a series of red trailing lights which meant the course was closed. The red lights were turned on in most cases when there was a sudden and definite wind change, which would make further landings on the previous course dangerous. In my own case, there were many incidents at night I had to take control of the plane or we would have made a crash landing. And, on some occasions, it was difficult to wrest control from the frightened student pilot. Even though night instructional duty was hazardous, we did not lose a single plane, but we did lose some planes and crews during daytime flights.

The next incident in my cockpit came on 6 October 1945. I had relieved LCdr. "Buzz" LaFever as Commanding Officer of VPB-108 in August at Iwo Jima, where we kept a detachment of six Privateers. We also kept a detachment of six planes at Peleliu, in the Palau Islands, and kept six at squadron headquarters on Tinian. During the period between the end of the war and early October, I was the only regular Navy officer in the squadron. All of the senior officers in the squad-

LCdr. Hill leaning on **Pistol Pakin Mama III**. 1st two were shot down.

Joe Hill, center standing, with crew before transpac at Camp Kearney, CA 1945.

Co-pilot Ens. Curtis beside **Mama III**.

Joe Hill, center standing, with crew after transpac at Kaneohe, TH July 10, 1945.

ron at the time the war ended were reservists, all had enough points to go home, and they were all gone by the 1st of October. This left me as squadron commander, with 17 patrol plane commanders who were Lieutenant (jg) with less than 1,000 hours of flight time, and very little of that as Patrol Plane Commanders in Privateers.

I was on Iwo Jima on 6 October with our detachment. The three detachments there rotated the duty every third day, and the duty squadron was assigned any special missions that day. About 0800 a message was received from the Wing Commander, and it was addressed to Commander Jamie Jones, Senior Officer Present. It read, "Assign pilot to locate and accurately report on a typhoon whose last known position was 150 miles northeast of Guam at 1600 on 4 October. Location this storm urgent in view of large number of ships transporting troops back to U.S." My squadron had the duty. Because my PPCs at Iwo had little experience in Privateers, I elected to take the mission. Our planes were not designed as weather planes, and in fact, they were ill-suited for such missions because twin 20 mm cannon had been installed in the nose of each aircraft,

The characteristic shape of the Beech 18, much used as a trainer in World War II and shown here in its first post-war version. (Beech)

making them extremely nose-heavy. I knew enough about storms and typhoons to know that I could locate the center of them by flying downwind with the wind on my port quarter. I took off at 0830 into a 30-knot wind blowing from the east-northeast, advising my navigator, Ensign Shoemaker, that I would be flying contact at 400 feet or less and to check with me every five minutes regarding wind drift and velocity. I had no idea where the storm was located, but I knew we would fly into the eye if we kept the wind on our tail in the 8 o'clock position.

The surface wind increased from 30 to 90 knots as we approached the center of the storm from the southwest quadrant. Flying conditions were rough, but not uncomfortable as we penetrated the eye. It was about 30 miles in diameter and the sun was shining brightly inside the wall clouds which towered above 50,000 feet along the north-east side of the eye. There were millions of birds being carried along inside the eye of the storm, because the wind was so strong outside of it that birds could not fly. I told my co-pilot, Ensign Curtis, to take over the controls and circle inside the eye, while I went back to check our navigation with Shoemaker. Our consensus at 1100 was that the center of the eye was 450 miles, bearing 220 degrees from Iwo Jima, and we asked the radioman to make this report to Wing Headquarters in Guam.

I went back to the pilot's seat and took over control from Curtis. The ocean's surface was riled and confused inside the eye, as there was no positive wind direction. Up to this point everything was fine, as there was absolutely no danger to the airplane or the crew of nine. However, the aerographer's mate 2nd class suggested that we exit the eye at the north-east quadrant, as this is the area with the strongest winds. I made the decision to try it, and this turned out to be the "sorriest" decision I ever made as a pilot in charge of a plane. We went from no wind to winds that were 130 to 150 knots, and this time the wind was slightly on our starboard quarter so that we could get away from the eye. For the next fifteen minutes, I frankly did not think we would ever get out of this situation alive, and as I looked over at co-pilot Curtis, I'm sure he did not think so either, for the cold sweat was running down his cheeks. The attitude of the plane had reached a frightful position in seconds after we entered the rolling base of the wall clouds. The nose-heavy plane with 130 knots of tailwind was in level flight with the nose down as much as 20 degrees. It was raining so hard and cooling down the cylinder head temperatures so much that the engines were running rough and coughing at cruise power settings. The rolling scud clouds were tossing the plane about like a piece of paper in a whirlwind. When we could see the ocean, it was a mass of white suds without waves. The wind was so strong that it was blowing the tops off waves and driving them into a surging wall of water more than 100 feet high at 5- to 10-mile intervals. Fifteen minutes of flying under such conditions does not sound like a long time, but when you feel that every second might be the end, it is several lifetimes.

We finally made it back to Iwo Jima after an 8.6-hour flight, and the Wing and Fleet Commanders had their fix (1100) on one of the most vicious typhoons of the year. This was apparent when the storm hit Okinawa with its full force on 12 October and literally devastated everything on the island except the native thatched huts which were protected by concentric hedgerows. As a result of my flight, I prevailed upon the Wing Commander to issue a Wing bulletin prohibiting penetration into the eye of typhoons. From that point on, typhoons would be located by circumnavigation at a distance of at least 50 miles.

My next incident in the cockpit came in October 1948. "Dixie" Harders and I flew to Reno one Saturday morning to get in flight time, but we were having so much fun gambling that we decided to RON (remain over night). Soon after arrival, I had bet on a horse that was running in the feature race at Bowie. The bet was $2 to win on "Cat's Meow." I thought no more about it until I returned to the horse parlor to see if I had won. I did not see the horse's name on the board, and I threw away my ticket in a cigarette sand box when I thought it had lost. Later in the afternoon we came back to the parlor to check some other results, and I learned that Cat's Meow had won and paid $18 for the $2 ticket. I began scratching around in sand boxes until I found my crunched ticket, and took it to the pay-off window and collected $18. "Dixie" and I thought this was such a good omen that we had better have a couple of drinks, and it was then we decided to send our wives a message that we would RON. We gambled until Sunday noon, and the total of our losses was less than $100, so we figured we had a worthwhile trip. By the time we got to the airport and filed our flight plan, it was 1 p.m. We had completed 80 percent of the flight and were flying along at 3,000 feet, about 500 feet above the Santa Cruz mountains, when both engines quit cold. We were both half asleep until this moment, and there was a mad scramble for the wobble pump and the gas tank selector valve, because we both knew what had happened. By the time both engines caught with a mighty roar we were almost into the tree tops. If we had been flying 100 feet lower, we would have bought the farm. There was absolutely no excuse for this type of pilot behavior while flying.

My final cockpit incident happened in a twin-engine Beechcraft while on an instrument flight from Atlanta, Georgia, to Washington, D.C., in November 1955. My co-pilot was Roy Suddath, and we cleared to fly at 8,000 feet, while reporting in by voice to all reporting stations. We had just about reached Roanoke, VA, when we began icing up so much that we could not maintain 8,000. I reported this to Roanoke, and requested permission to take up an easterly heading while we were losing altitude, hoping we would get warmer air near the ocean. This heading was also away from the mountain ranges. Roanoke gave us permission to take a heading of 090, and asked us to advise when we could maintain level flight. We were down to 1,000 feet and almost in the Norfolk control area before we broke out of the overcast and reached a maintenance point. We advised Norfolk approach control that we were on visual flight rules and no longer had an icing problem. We asked them to advise Roanoke that we were proceeding VFR to NAS Norfolk to land and refuel. This was a case where weather conditions were such that we should never have been cleared for instrument flight to Washington in that type of plane under such severe icing conditions.

Before proceeding on to individual accounts of cockpit incidents, I think it would be well to make some general observations. Any pilot who has ever had a near miss with another plane in the air qualifies and so does any pilot who has had to make an emergency landing, or to ditch a plane at sea. Any pilot who has engaged an enemy plane in combat, or dropped bombs on enemy ships qualifies. Any pilot who has had one or more engines fail in flight, or had a fire in a plane qualifies. And finally, any pilot who has had to bail-out of a plane certainly qualifies as a survivor of a cockpit incident.

Harry Ginn created an incident at Pensacola when he almost chewed up my tail and cockpit while we were shooting circles, and this very nearly cost the lives of two cadet students and the loss of two "yellow perils."

Classmate Jack Grayson was shot down twice in the Pacific. The first time while he was on a bombing raid to Menado in the Celebtes Islands in a PBY. He spent seven days adrift in a rubber life raft, before he made it ashore on the small island of Sanara. He was picked up a few dsys later by classmate Jack Davis in a PBY. He was shot down the second time while on a photo mission over Wotja Island in a four-engine Privateer and had to crash-land on Kwajalein. These were the only two incidents mentioned by Jack in his correspondence, but I'm sure that he and his shipmates had several in the cockpit while their squadron was retreating from the Philippines in 1941-42.

Classmate Arbuckle was the only survivor of his eleven-man crew when he was shot down in the Pacific and taken prisoner.

F4B-4.

Classmate DeLorenzo says the hair was standing on his bald head when he was making a bombing raid on Wake Island after the Japanese captured it. He was flying a four-engined Coronado flying-boat and had been advised by intelligence officers to keep his engines out of synchronization while making his bombing approach, because the accuracy of Japanese A/A fire depended upon whether or not the engines were synchronized. Delo thought this was silly until the A/A burst began bouncing his plane all over the sky — he changed the throttle settings.

Classmates Coley and Corlett landed at Espiritu

Santo among a harbor full of darkened ships with live torpedoes about midnight 13 November 1942, after our torpedo strike mission had been called off.

Squadronmate Clark made two PBY deadstick landings in the open ocean — the first at about midnight when he and Willis missed the islands after a 1,500-mile war patrol out of Kaneohe. They ran out of gas, and Clark made a perfect landing in the 10-foot seas without any damage to the aircraft. The second deadstick landing was made when Japanese Zeroes shot out both engines during his vertical dive, as mentioned earlier.

Squadronmate Willis was shot down by a Japanese four-engined patrol plane. They were picked up by one of our PBYs on 10 September, after almost four days in two life rafts.

Two other squadronmates, Howe and Reese, were killed along with their crewmen when their planes crashed during bad weather.

Squadronmate Ed Kiem had two incidents while in training at Pensacola, the first when he was making solo practice landings early one morning at K field. On his third landing, he gave the plane full throttle after his wheels touched down and his engine quit cold. He applied his brakes as he was skidding on the heavy dew; but he reached the fence at the edge of the field before he stopped, and the plane flipped over on its back. Ed was not injured until he unbuckled his safety belt, fell to the ground on his head, then was hit in the head by the fire extinguisher which he had tried to grab on the way down. Even so, his injury was not serious, and he rode back to Corry Field with the crash crew. His second incident happened at night after a cross-country flight. Fog had rolled in over Corry, and most pilots in the flight flew back west and landed at fields that were clear. But Ed's girl friend, Harriet, was waiting for him to take her on a date, and he considered it urgent that he land at Corry. He could make out the red obstruction light of the tower so he came over it at 50 feet and chopped his throttle. He made a hard landing with a few bounces in the dense fog, but there was no damage to either the plane or to Ed, so he carried out his date with Harriet, who later became his wife.

Charlie Harbert, 107 C, ejected himself from an F4B-4 near Corry Field one night flight when he found the field socked in by fog. He landed in the Escambia River, as related earlier.

Jack Poteat's, (150 C) first incident came in May 1941 after a 20-hour search flight for the *Bismarck* out of Argentia, Newfoundland. The bay was completely socked in by fog, and he made a totally blind let down and landing at night. During the Battle of Midway, when PBY pilots were told not to land at Midway in view of imminent enemy attacks, Jack had a B-17 pull up beside him and asked for a bearing and distance to Midway. Jack escorted the lost pilot back to Midway. He then proceeded on to French Frigate Shoals to land. Jack and his crew flew 174 hours that month, mostly in the early part of June during the Battle of Midway. His next scare came in the South Pacific on what was to be a routine four-hour flight between Noumea and Espiritu Santo. It was 8 January 1943 and about 150 miles south of Noumea where he encountered gale force winds and the weather deteriorated into a full blown typhoon with winds of 100 knots or more. It took him twelve hours to make the 450-mile flight. About a month and one-half later Jack sank a Japanese submarine that he caught on the surface about 150 miles northwest of Espiritu. After the war, Jack became an airline check pilot for the FAA. During his 27 years in this capacity he had two incidents where he had to take over control of fully loaded passenger planes from pilots he was checking or the planes would have crashed. Jack never would tell me which airlines were involved, or what happened to the pilots in each case.

One of the first aviation cadets to get his wings at Pensacola was Walter Hibbs in Class 81 C, who served two years in the Fleet before getting orders to Seattle to serve the final year of his four-year contract as a cadet. He completed his active duty in July 1939 and flew with Pan American Airways until he retired at the mandatory age of 60 in 1972. Once on a Boeing Stratocruiser (B377) on a PanAm flight between Hawaii and San Francisco, it was a moonless, clear night and they were about half way to San Francisco when number 3 engine faltered. Walt had his hand on the throttles and was about ready to feather the propeller on number 3, when there was a sudden jolt and the instruments went dead. There was also a burst of flame that lighted up the entire cockpit, as the engine caught fire, trailing a sheet of flame back to the tail section. As the co-pilot began shouting "Mayday" on the radio, Walt picked up the microphone and awakened the passengers with the alarming news that number 3 engine was on fire and they might have to ditch. One little old widow, whose husband had been a Navy flier killed in the war, came up the stairs from her bunk and ran screaming down the aisle shouting, "We're all going to die." She had to be subdued by other passengers. The engine fire was out within a minute, as Hibbs leveled out and began checking everything. Number 3 engine was dead, and the prop was feathered, but there was no problem in maintaining level flight, so Hibbs went back into the cabin to soothe the passengers. In the back of the cabin he found the purser trying to console the widow. Hibbs invited her to join him in the cockpit, where he sat her in the jump seat. She listened intently as he explained all emergency procedures and the airline's safety record; she then was calm enough to go back in the cabin to try to get some sleep. He bid her goodnight and invited her back in the morning to watch the beautiful sunrise. It was a long night as Hibbs and the co-pilot took turns cat-napping. The engineer had his seat back as far as it would go with one eye half open, and the navigator was sleeping between fixes with his head in his hands at his table. The sun was coming up over the horizon and the dozing Hibbs felt a tug on his shoulder as the widow got his attention. Before he could speak she said brightly, "You know, Captain, this is the most reassuring thing that has happened all night. If you people don't take things any more seriously than this, I guess I shouldn't either,'' and she left the cockpit.

C. O. Rogers, Class 83 C, did not have any incidents in his cockpit, but reported some that happened to others. Gerald L. Pentz, a classmate of Rogers, was on a night-flying solo flight in Squadron Three when he became disoriented and decided it was time to leave the airplane. He had plenty of altitude when he bailed out, but was jerked violently when the parachute opened, causing the flashlight he was wearing around his neck to deliver a sharp blow to the head which knocked him unconscious. He was not the first cadet to come down in an airplane in an unconscious state, but he was the first one to come down unconsciously in a parachute. He did regain consciousness long before he hit the ground, landed safely, and became the first cadet to qualify for membership in the Caterpillar Club.

Perhaps one of the most famous cadets in the early classes was Bob Stanley. Although I do not have any incidents reported while he was in the cockpit, I'm sure he had many of them. After serving out his cadet contract and receiving his $1,500, bonus he became a test pilot for Chance Vought. He then became a Chief Test pilot for Bell Aircraft, where he designed and tested the Aircobra XS-1. He later formed and managed his own aircraft manufacturing company in Buffalo, NY, and finally moved to Stapleton Field in Denver where he designed and manufactured ejection seats and other safety equipment. Bob made flying safer for all pilots who followed him, because of his professionalism as a test pilot.

Robert George Dose, Class 94 C, was designated Naval Aviator #5333 and did not report any incidents, but he did recall a number of firsts which may have made the hair rise. He was a member of the first F4U-1 Corsair squadron aboard the *Saratoga* in World War II, and shot down two Japanese Zeroes. He later became Commanding Officer of the squadron. He made the first mirror landing without a landing signal officer aboard a carrier flying an FJ-1. He fired the first Fleet Sidewinder missile from an F9F-8. He made the first carrier-to-carrier non-stop flight across the U.S. flying an F8U-1 Crusader in the record time of 3 hours and 28 minutes, which still may be a record. He made the first carrier landing in the F3H-1 Demon and the first in the F8U-1 Crusader. I served with Bob aboard the USS *Princeton* in 1946 and 1947, and I can honestly say that he was the finest Air Group Commander I have ever known. Furthermore, along with his classmate, Walter L. Curtis (Air Officer of the *Princeton*), he maintained the finest air department-air group relationship I have ever seen on a carrier. They have remained fast friends since they were cadets in 94 C in 1936 and are both retired in LaJolla, California.

Walter C. Michaels, Class 97 C, had his first incident at Pensacola when he was not even in the cockpit. He forgot a parachute and got 100 demerits. The next incident for Mike came during multi-engine flight training at Jacksonville. One of the B-24s was lost and Mike and his crew were ordered to initiate a search pattern for it. The seas were rough and visibility was limited in the low overcast, and it was a completely dark night with no horizon. Mike and his crew spotted a bobbing light on the surface and went down to take a look in case it was a life raft. After the pass, Mike pulled up into a sharp left turn in order to keep the light in sight, but when he checked his instruments he found that he was in a 60 degree left turn and losing altitude rapidly. With great effort he leveled his wings and felt like he was still in a tight left turn. He was almost in the water before he checked his loss of altitude and began climbing to 3,000 feet. He gave his co-pilot hell for allowing the plane to get in such an attitude, but it was strictly a case of vertigo for Mike and he seriously thought about turning

A hoped for sight! A U-boat on the surface at 1330 in the afternoon. (Civil Air Patrol)

in his wings for such carelessness that nearly cost the lives of his crewmen and the loss of a B-24.

Squadronmate George Enloe was my co-pilot during the Battle of Midway and endured one of his incidents when we landed our PBY-5A with parking brakes on and blew a tire. In another instance, George made contact with a Japanese submarine on the surface and radioed that "He refuses to submerge so I can attack him." He then tried a high-level run and missed. On another occasion, while flying patrols out of Ndeni and Gracioso Bay, he damaged the hull of his plane so much on takeoff that he knew a safe landing was improbable. Nevertheless, he elected to complete his 1,500-mile patrol before attempting a landing. He made a fine landing near the end of the Bay, but the plane was taking on so much water as he was taxiing toward a buoy that the plane was sinking nose first. The pilots from all three squadrons on the tender were cheering for George. He suddenly realized that a sunken plane moored to a buoy was no good, so he headed for the beach. He was still 100 yards short and the propellers of the Pratt & Whitneys were continued turning over as George's plane went completely under nose first. After the war, George flew thousands of hours as an airline Captain, but unfortunately was killed while riding as a passenger on the Eastern Airline flight that went into the Everglades several years ago.

Squadron and roommate George Poulos also had a few incidents. The first came during the first month of the war when George was operating from the *McFarland* in Hilo Bay and was ordered to locate two PBYs from the States which had missed the islands and were down at sea. This he did with coughing engines, as narrated earlier. George did make a safe landing eventually, but ran out of gas before taxiing to the beaching ramp. The second incident for George and his crew came while the squadron was flying war patrols from our newly constructed base at Saweni Bay in the Fiji islands. On these patrols we often flew over small volcanic islands which we inspected closely to see if there was any evidence of Japanese landings. This gave our pilots a feeble attempt at "flat-hatting," and also gave them a glance at the topless native women. On one such pass George Poulos made a run along the beach at about five feet above the water. One old native warrior stood his ground in the surf and threw his spear at the big ugly bird which he thought was invading his island. The spear actually ripped a hole in the rudder of George's plane, and he felt the jolt as it hit. It did not affect the flight characteristics of the plane, however, and George continued his patrol.

Jim "Andy" Hardy, Class 132 C, helped form VP-63, the first squadron to develop and use MAD (Magnetic Anomoly Detection) equipment. The squadron was formed in 1942 at Alameda, and two of his shipmates in the new squadron were classmates of mine in 130 C — Jim Breeding and Woody Shoals. Unfortunately, both of them were killed before the squadron's first tour ended. Jim Breeding was lost during a severe storm off the Oregon coast on New Year's Eve 1942, and Woody Shoals was killed while taking off in a plane at night at Port Lyautey, North Africa, less than a year later. The MAD equipment was designed to locate and kill submerged submarines, and the squadron's first deployment was flying out of an RAF base, Pembroke Dock, in Wales, to patrol for submarines out of the Bay of Biscay. Tides at Pembroke Dock sometimes varied as much as 20 or 30 feet, and this made PBY beaching hazardous. One day when one of the VP-63 planes was making an approach that would surely wipe out a wing-tip float, both wing-tip floats were retracted while the crewman on the wing stepped to the port wing allowing the starboard wing to miss the ramp. Then the wing-tip floats were lowered to give the pilot a chance to make a decent approach. The British and Australian flight crews marveled at the "bloody skill of the crazy Yanks." Little did they realize that the whole thing was an accident, as someone tripped the wing-tip float switch at the right moment, and then, realizing what he had done, put the same switch in the down position in the nick of time. The squadron lost some planes to German JU-88s, including that of Bill Tanner who had sunk the Japanese mini-sub at Pearl Harbor before the main attacks on 7 December. However, Tanner and his crew were rescued from the Bay of Biscay without injury.

Gordon Egbert, Class 107 C, had his first incident on his first dual instruction flight as a Seaman 2nd Class. He forgot to fasten his seabelt, and when his flight instructor, Harry Sartorris, was at the top of his loop, Gordon was hanging on for dear life. He never got in a cockpit again without fastening his seat belt and shoulder straps. The only other incident he reported was when he was flying night anti-sub patrols out of Key West after the war started. He had 91 octane gas in one tank and 100 octane gas in the other, as he used 91 for cruising and 100 when making an attack or inspecting anything low on the water. The weather was bad, with a 60-knot gale blowing in moderate rain, when he spotted something on the surface. He asked the mech in the tower (the pillar between the hull and the wing of a PBY) to switch to 100 octane and went down to investigate. The object in the water turned out to be a merchant ship rather than a submarine, so he climbed back up to 1,400 feet and told the mech to switch back to 91 octane, asking how much fuel remained in that tank. The mech's response was "200 gallons." Ten minutes later both engines quit, and they were down to less than 400 feet by the time the mech switched to the 100 octane crossover valve and the engines caught again. Gordon only lost one other engine during his career and that was in a B-24 while flying out of Hutchinson, Kansas. That presented no problem for him as he had the field in sight and came on in for a safe landing. Gordon's only other incident was a happy one. He and his good friend Roger Crowley both commanded Privateer squadrons in the Philippines. Gordon had VPB 111 in Palawan, and Roger had VPB 106 in

Mindoro. Gordon and Rog had a lot of fun with French names while training Free-French students at Jacksonville in 1942 and 1943, so one day when Gordon was on patrol over Hinan Island, he opened up on voice radio with "Hello Gebert, Hello Gebert, this is Gecoquile — you hear me, no, yes, over." Back came the response, "This is Gebert, This is Gebert, you follow me back to base, yes." So Gordon followed Rog back to Mindoro, and "spent" the night as they polished off Roger's liquor ration. Gordon flew back to Palawan the next day.

Art Farwell, also in Class 107 C, wiped out the undercarriage of an F4B-4 while "skipping rope" (simulating carrier landings) in Squadron Five, but the board let him off when his instructor admitted he was giving him the "too fast" signal when he spun in! In June 1942 his squadron, VP-84, was flying out of Argentia, Newfoundland, and Reykjavik, Iceland, in PBY-5As. Art was flying off the coast of Norway one day covering a Murmansk convoy when he was jumped by a German Focke-Wulfe 200, a four-engine bomber with lots of fire power. He escaped the German plane by diving into a fog bank and flying on instruments at 100 feet until they were clear of the area. On another occasion he made a hairy landing at Reykjavik when visibility was bad and the runway was completely iced over. He managed to land OK, but the plane began a slow ground loop until he was 180 degrees from his landing direction, and then he gunned both engines and brought the plane to a safe stop.

His next incident came in the spring of 1945, while he was Executive Officer of VPB-118 flying out of Tinian. His squadron planes would fly to the coast of Japan to destroy small shipping then return to Iwo Jima, refuel, and return to Tinian. On one of these missions, Art had strafed a small picket vessel and had sustained some wing damage. Even though he had refueled at Iwo, his wing tanks were leaking so much that he did not think he would reach Tinian. He had a crew of 13. It was Friday, 13 May. He was at 700 feet altitude, five miles from Tinian, when the engines began to stop from loss of fuel. His crew was already at ditching stations, and Art made a smooth dead-stick landing, but they could only get one life raft out of the plane before it sank. There were eight men in the raft and five hanging from the side, including Art. He made the remark to his crew, "My first command at sea, and I can't even get in the damn boat." They were picked up by rescue boat in 32 minutes.

That same May, Art and a wingman in another Privateer spotted a tanker and a destroyer escort in a rather restricted area of a channel near Seoul, Korea. Art's wingman made a feint at the escort while Art popped up over a small island and dropped his bombs on the stern of the tanker, making a direct hit. But before he could close the bomb-bay doors, a three-inch shell from the escort exploded underneath his starboard wing. His number 3 engine caught fire, and the number 4 engine had a big hole in it and was spewing oil. The co-pilot feathered number 3, hit the fire extinguisher, and put out the fire, but the rpm on number 4 was rising rapidly as the engine ran out of oil and it, too, had to be feathered. With engines 1 and 2 at full throttle, the bomb-bay doors open, and no flap control, Art could not maintain flight, and for the second time in the month he had to ditch a plane. (Again they had a crew of 13). It was a hard landing this time, and the plane broke into pieces and sank rapidly. The impact of the hard landing had inflated one life raft, trapping a radioman at his station, and so a buddy shot holes in it to free him. Art had a shrapnel hole in his life jacket and had to grab a cushion. Only one four-man life raft was inflated, and the bow gunner and one waist gunner, both wounded by shrapnel, were hoisted aboard, with two men to comfort them. The remaining nine men were in life jackets or hanging together in the water, but they looked even worse when a Japanese plane started to strafe them and a Japanese destroyer appeared on the horizon approaching them at flank speed. However, their wingman was not standing idly by, as he called in two Privateers on prowl nearby, and they in turn notified two Mariner seaplanes about 50 miles away. The three Privateers turned away the destroyer and the strafing Japanese plane as the two Mariners came to the resuce. One Mariner landed near the men in the water and hauled them in the plane through the waist hatch. Art was the last man in, and the pilot started his takeoff before he was completely in the plane. Art and his crew got credit for 3½ Japanese planes shot down (he and his wingman shot down one simultaneously), but he never again flew with a 13-man crew after the second ditching.

Harvey Hop, Class 133 C, found that many of their VP-24 planes were lost on the water due to Japanese bombing and strafing attacks on 2 February 1942 at Sourabaya, in the Dutch East Indies. After the big attack and the loss of his plane, Harvey and crew were sent out on patrol in a Dutch PBY and ordered to operate from the seaplane-tender *W. B. Preston* on the south coast of Flores. After three days of negative patrols, he and Guy Howard were called in for a special mission with Guy acting as the PPC. The mission called for a night, open sea, takeoff and a flight to the island of Banda where they were to pick up some political prisoners and return them to Flores. Harvey says that night takeoff in the open sea was the most terrifying incident that happened to him in 37 years of flying. Huge swells were running that night, as Guy elected to take off into the wind and into the swells. Harvey's instructions as co-pilot were to "two-block" the throttles if the plane got airborne and try to keep it airborne. As the plane began to pick up speed and get on the step, it started a series of crash bounces as it went from the crest of one swell to the other. After about 15 such bounces, they finally became airborne and flew to Banda where they landed at dawn. Guy went ashore in a boat that met the plane, and they came back in 45 minutes with two men and two women. One of the men was Sukarno, who was

The 1,000th PBM Mariner built by Martin takes off from the waters of Chesapeake Bay. (Martin-Marietta)

later to become President for many years. The takeoff from Banda was a little hairy, as it required getting on the step going down a gently curving channel, then making a rather sharp turn through the trees once they were airborne; but the flight back to Sourabaya was uneventful. Harvey and his crew shot down two Bettys and one Zero during the South Pacific tour.

Another incident happened to one of the planes while they were at Sourabaya. One captain, an Academy man, and his crew were flying patrol at 10,000 feet when they spotted a formation of enemy bombers. He put his plane in such a violent dive that he ripped the fabric from the trailing edge of the wings before he leveled out at less than 1,000 feet, where he ordered the navigator to bring him a parachute. The captain did the impossible after he donned the chute, as he went out the navigation hatch between the propellers and bailed out. An enlisted pilot was getting ready to bail out when he noticed that most of the crew in the after hatches were entangled in their parachutes as a result of the violent dive and could not bail out. He went back to the pilot's seat, and he and the navigator brought the plane in for a hot open-sea landing. They all got out safely, but the enlisted pilot was seriously wounded by the strafing bombers after they were on the water. About one week later the captain said he was ready for another patrol, but when he reached the plane in preparation for the flight, there was no crew there to fly with him, and so his mission was cancelled.

Hamilton Lawrence, a Marine cadet in Class 107 C, had his hairiest flight years after the war, in Switzerland, where he was chief pilot for a Lear Jet charter operation with 14 pilots under him. On this occasion he had a flight into Innsbruck, Austria, to pick up a lady stretcher patient who had a stroke and was to be returned to Antwerp, Belgium. The weather was bad and deteriorating rapidly as he approached the only navigation aid atop a 14,000-foot mountain above the town and airport nestled on the valley floor. He circled the marker beacon at a safe altitude above the mountains until he spotted the runway through a break in the clouds. There was no instrument procedure, so he dumped the speed brakes and put down the landing gear and dove for the runway. He got under the cloud cover and landed without difficulty. While he waited for the patient and filed for Antwerp, it began snowing. Visibility dropped to 200 feet, and the ceiling was about the height of the terminal building. A Swissair plane inbound had to return to Zurich. Climbing out blind from an Alp valley more than two miles deep, without any navigational aids, did not present a very attractive problem, but he felt obliged to get the patient to Antwerp. He planned to hold his takeoff heading for 45 seconds, then turn left 10 degrees, which would allow him to climb out in the widest part of the valley at 7,000 feet per minute. He never did break out of the clouds, but when he passed 14,000 feet, the height of the mountain tops, he knew he had it made. He delivered the patient safely to Antwerp and medical help.

Don Munro, Class 154 C, was given a 20-hour flight check instead of a solo flight check, and that was a hairy ride, as he was afraid to side-slip into small fields. Don had heard that this was a no-no on a solo check, so he made a series of S turns into the field each time. Finally, Captain Nevils, the Marine check pilot, asked if this was a solo check. When Don nodded, yes, they went to 4,000 feet and spent the last part of the hour wringing out the plane. Don got an up-check. On Don's first night solo flight, he was making his second landing at Corry when he observed another solo pilot making a short circle. He could see red and green running lights coming right at him, so he dove the plane about 200 feet to avoid what would have been a fatal crash for both pilots. Don also reported a fellow classmate who had received Commandant's time and had one up and one down. He needed only one deadstick landing to the 100-foot circle for another up. He got the plane in the middle of the circle, but stalled it in from 50 feet and washed the plane and himself out of the program. Don had a couple of other hairy flights that were caused by bad weather. On one he iced up on a flight from Oakland to Seattle, and had to return to Oakland. On the other, on 18 July 1944, he was flying 25 passengers from Finschaven to Milne Bay, Australia, and the ceiling and visibility at Milne Bay was under instrument minimums. He let down the first time to the tree tops, but could not find the strip and had to climb out and make another pass. However, on the second pass, he broke out at 100 feet right over the runway and landed.

Marine cadet Harry Hooper, Class 146 C, had a couple of amusing incidents at Pensacola. The first was seeing a big husky cadet, a former football player, walking in barefoot from the flight line carrying two parachutes. He was stepping very gingerly on the hot asphalt, as his instructor was trying to teach him to have a lighter touch on the rudder pedals. They had almost lost the plane during an emergency, when the instructor had a difficult time gaining rudder control from the student. The other incident did not happen to a cadet, but happened directly in front of them when they were lined up for muster and morning roll call. The Skipper of Squadron One, Cdr. "Dusty" Rhodes, always flew an N3N from mainside to Saufley Field, and this morning was no different except for the landing result. Dusty ignored the usual flight pattern and approach procedure and landed on the taxiway in front of the mustered cadets. He came in a little hot, and the plane nosed over on its back when he tried to stop it. Embarrassing, to say the least!

After receiving his wings, Harry Hooper was retained at Pensacola as a flight instructor. One day he was riding in the rear seat of an SBC-4 with student Buster Hoyle in the front seat as part of a nine-plane formation doing cross-overs. The entire formation was led into a thunderhead at 6,000 feet, and it was every man for himself. The rain was coming down in torrents. In fact, Harry was getting very wet in the rear seat as he had to keep closing the canopy as fast as Buster was opening it. Harry did not learn until they landed that Buster was trying to bail out all of the way down to 2,000 feet where they broke out of the turbulent thunderhead.

Don Kitchen, of 6A41-J at Jacksonville, reports that a cadet in his class was flying solo in an N2S at 2,000 feet and could not recover control from a spin. Finally, in desperation, he hauled back on the stick just before he crashed in a clear area. Don said that the only part of the airplane that was recognizable was the rear cockpit. The student walked away from his crash unhurt and reported it to his flight instructor who was coming down the road looking for the crash site. Don says they needed pilots pretty badly because they allowed this cadet to finish the course and get his wings. Don reported another incident at Jacksonville, where a cadet started a plane with no one in the cockpit. The situation was just about out of control before some fellow cadets came to his rescue and one of them managed to get into the cockpit and shut it off!

Billy Alford, another cadet in 6A41-J at Jacksonville, reports he had completed his last instruction in night flying with Ensign R. D. Bonnell, and he was ready for his night solo flight in an N3N on floats. However, Bonnell, Class 135 C, suggested that Alford change planes because the engine had been running rough on their flight. Billy changed planes and taxiied away from the ramp out into the St. Johns River and took off. Once he was in the air, he eased back on the throttle and began to check his instruments. His airspeed was O--the mech had not removed the pitot tube cover before the plane was launched! Billy got the Very pistol and began to shoot flares down toward the river to indicate he was in trouble, but not one flare worked! Billy seriously thought of climbing to 2,000 feet and bailing out, but decided to try to bring the plane down on the water. After a series of attempts, he finally got the aircraft back on the river, well away from the other planes that were making touch and go landings, and taxiied back to the ramp. That for him was enough of flying "by the seat of his pants" for one evening.

Early flying was done in the U0, the landplane version of Consolidated's NY-2. (Naval Aviation Museum)

Classmate Norm Brady and his crew were sent on an all-night tracking mission on 26 October 1942 when a large Japanese task force was sighted in an area east of Guadalcanal. There was heavy cloud cover that night, and the first indication Norm had that they had contacted the enemy force was when they flew directly over the middle of it at 1,000 feet and the ships' searchlights began sweeping the skies. Norm made his contact report, then headed for the nearest island to make a navigational check. This heading took him right over an advanced force of two cruisers and three destroyers heading directly for Ndeni where our planes were operating from the tenders *Ballard* and *Mackinac*. Norm sent a message to the tenders with information ComAirSoPac, and the ships were able to get well clear of Graciosa Bay before the attacking ships reached the harbor. Also the planes moored in the Bay were ordered to fly to Vanikoro, for the next day's operations. Norm's action probably saved the tenders and the PBYs and their crews from destruction. There were several VP-11 pilots and crews there that night, including myself. Squadronmate Ed Kiem was on patrol the next day, spotted the advanced force, and was chased by several Rufes. In the excitement Dudley Dexter, the navigator, stepped through the bombardier's window as he went to man the bow gun. Ed said he thought the plane had been hit, as the fuselage became a wind tunnel. There were no tenders at Ndeni when they came back to land that night, and they took on tons of water when they landed with no window in the bow. They bailed all night and took off at first light for Vanikoro, with tons of water after takeoff and tons of water after landing.

Charlie Kinsella, Class 108 C, was involved in a night squadron formation exercise one night off Virginia Beach. There were two divisions of nine planes each, and the flight was taken out over the Atlantic. It was a dark overcast night, and all Charlie could see was a bunch of bobbing running and formation lights mixed with the flickering flames of exhaust stacks. All of a sudden Charlie got the feeling that he was in a steep climbing turn, and it was all he could do to stay in formation. The urge to break away and save himself was almost overpowering until they were headed back over the beach and he had a real horizon once again, thus negating his attack of vertigo. Another incident happened to his "bunkie" on the *Ranger* in the same squadron with SB2Us. The plane his "bunkie" was flying had undergone a 120-hour check the day before and had been spotted on the flight deck in an "up" status for launch, but the minute he got in the air he discovered that his ailerons were crossed. If he moved his stick to the left, his right wing went down, and the reverse was true when he moved it to the right. He aborted the mission and came back to land remembering the LSO was "crossed-up." Charlie also reported he was taking off in a single-engine amphibian J2F one day at NAS San Juan, PR and the engine quit cold when he was about 150 feet in the air. He did not have enough runway to make a landing on the strip, so he cranked up the wheels and made a deadstick landing in the water. He had taken off with the fuel selector on right auxiliary instead of on one of the main tanks. It was pilot error, but no damage was done. Charlie also described how he was flying from Guantanamo to Miami VFR one day when conditions went to instruments, and the R50 went into a graveyard spiral. Charlie was the co-pilot, but as the airspeed built up and the sliding Plexiglas panel on the pilot's side blew out, he pushed the mixture controls to rich and took over the controls and leveled the wings. They broke out of the overcast at 1,000 feet and completed the flight on visual flight rules.

Tom Sutherland, Class 119 C, had a couple of incidents while he was in a patrol squadron down in Panama. The first one happened when he was trying to land near Oak Ridge Island to see if he could learn anything about refueling operations for Nazi submarines. He was coming into about 15 knots of wind for a landing, heading directly for the island. He had already committed himself to land when the wind suddenly shifted 180 degrees, so that he landed downwind very hot, and this carried him and the plane right up on the beach into a thicket of small bushes and trees. The propeller blades were all bent back about 90 degrees about one third of the way to the hubs, and tree limbs had penetrated the hull on both sides just below the wing spars. Tom surveyed the situation along with one of the native islanders, Tom pointed out to him that he needed a saw, a hammer, and a block of wood. He used the hammer and block of wood to straighten the propeller blades, and he sawed off the tree limbs on the outside of the hull. He also asked the islander to get a motor boat and a tow line to help get the plane back in the water. When the native came back with the boat, he handed Tom a bottle of rum and told the crew not to worry, as he would get the plane back in the water — and he did, within thirty minutes. Tom took off and flew the plane back to Panama. He never did find out anything about the Nazi refueling operations.

Tom's second incident came shortly after the war began, and it was much more serious. It was January 1942 and he and his crew had to land their PBY in mountainous seas (waves 30 to 40 feet) because of bad gas. The hard landing buckled the fuselage from the cockpit bulkhead to the waist hatch bulkhead, but the navigator and two radiomen escaped through the navigator's hatch unharmed, and none of the crew was injured. The central part of the plane was completely awash due to the buckled hull, and they were fearful that the plane would break apart and sink, so they used the anchor line to tie the tail section to the 50 cal gun barrels in each waist hatch. In order to stay afloat, several crew members had to bail constantly. They had lost all of the food except two cans of fruit, and they had no fresh water. They were hundreds of miles outside any shipping lanes on the Pacific side of Panama, and they had no communications equipment that would work. Tom knew their situation was desperate, and he insisted that they all get down on their knees and pray periodically, because he had always been taught that "God helps those who help themselves."

About the second day the wind died down somewhat, and they began to get some of the radio transmission equipment out of the plane and onto the wing to dry. They salvaged a gasoline stove from the plane captain's station, and they were able to get the auxiliary power unit out of the tail section. They used the gas stove to heat sea water and made a crude evaporator out of tubes inserted in the fruit can which condensed a small amount of fresh water from the steam. This allowed each man about a tablespoon of water per day. There was no rainfall during the 7½-day ordeal, and they had no food except for the fruit they divided from the two cans. About the fourth day they were able to make CW Morse code transmissions in the blind, but they had no receiving equipment. One of these transmissions was picked up by a ham radio operator in Pennsylvania, who relayed it to the message center in Norfolk, VA. Norfolk in turn relayed the message to Panama (who had given up searching for Tom and his crew), which passed it to a small task force of two cruisers and three destroyers which were on the Pacific side of Panama and were the closest ships to the downed position Tom's radioman had been sending in the blind. On the sixth day they got a radio receiver working, so they heard the radioman from the cruiser *Trenton* requesting they send MOs for five minutes at the beginning of each hour. They told the *Trenton* radio operator that they would transmit the exact time of sundown. From this the *Trenton's* navigator worked out a line of position, and the following day an SOC from the *Trenton* came out and found them, but as he was at the maximum distance of his search, he did not have enough fuel to circle the downed plane and let them know that he had seen them. They were picked up by the *Trenton* late that afternoon. The waves were so high that Tom stepped from a whale boat onto the deck of the *Trenton* which had a 20-foot freeboard. Tom and his crew had the greatest respect for each other, because no one had panicked, and they all knew that the good Lord had brought them through safely.

Chapter Eight
Where They Are Now

Most of the cadet-trained Marine and Naval aviators who finished flight training before World War II are now in the 65- to 75-year age bracket. Many of them have retired from business or active employment, but some are still flying or are working as consultants with aviation companies; others have become writers and authors. Some are lawyers, judges, doctors, or other professional types. Some are entrepreneurs, and some find it is a full-time job just looking after their investments. The stories that have been sent to me represent only about ten percent of the cadet-trained pilots who are still living around the world, but the ones I received all point toward one conclusion. Whatever these fellows attempted, they did a fine job at it, and in almost every case they were highly successful. During my flying career I got to know a large number of these pilots personally and have first-hand information about what they did when they were in the service. I have learned a great deal from their letters about what they have done since then.

After World War II and the end of my tour as Commanding Officer of VPB 108 in Okinawa, I returned to the U.S. and went to Damage Control School in San Francisco to prepare myself for a tour of duty on the carrier *Princeton* (CV-37). I was ordered to the ship on 1 July 1946, and her first mission was to deliver President Manuel Quezon's body back to Manila, as he had died while on an official visit to the U.S. I was Assistant Air Operations Officer, and had the pleasure of working for one of the finest Air Officers who ever served aboard a carrier, Walt L. Curtis, Class 94 C, and working with the Air Group Commander, Bob Dose, also from 94 C, one of the finest Air Group Commanders in the business. They had both been shipmates on the old *Princeton* (CVL-23) with Capt. John Hoskins when their ship was sunk by Kamikazes during the war, and they wished to be a part of his team when he took command of the new *Princeton* (CV-37). This was the second carrier that Walt had been aboard when the ships were sunk in the Pacific, as he was the Landing Signal Officer of the *Hornet* when it was sunk. Although Walt was a two-time loser while he was aboard

The standard USN carrier of World War II — the second ***Yorktown*** *at Pearl Harbor in September 1943 while on a ferrying mission with a deckload of Army jeeps and two Venturas. (Clark Reynolds — Navy Department & National Archives)*

Captain George J. Dufek, USN.

carriers that were sunk, he went on to achieve the highest rank of any cadet-trained Naval officer and is a retired Vice Admiral living in LaJolla, California.

The fine training I received aboard the *Princeton* prepared me for a tour of duty aboard the USS *Antietam* (CV-36) as Air Officer during the Korean War. I was a student at the Armed Forces Staff College in Norfolk when that war started and was ordered to the *Antietam* to assist in bringing it out of mothballs and preparing it for Korean fighting. As Air Officer, I was fortunate in having a cadet-trained pilot with much carrier experience as Operations Officer. He was Charles E. Roemer, Class 105 C, and his invaluable carrier experience was a great help to me, as I had never landed a plane aboard a carrier and was probably the only Air Officer ever sent to war without being qualified in carrier landings. I was also fortunate in that the four commanding officers of the primary attack and fighter squadrons were all cadet-trained pilots with World War II experience on carriers. After the war they remained in the ready reserve as weekend warriors and had at least two weeks of active duty as pilots each year, operating as complete squadrons. The squadrons were from different geographic locations in the U.S. and were called into service for the Korean War with all new airplanes. The COs were all Lt. Commanders. Roland Kenton and Tony Denman had the two jet F9F fighter squadrons from New York, and Bob Doering had the F4U fighter squadron from Denver. Buster Bitting had the AD attack squadron from Chicago. The Air Group Commander was Bob Farrington, the only Naval Academy graduate aboard except George Dufek, the Captain, and Jim Newell, the Executive Officer. Because I had no carrier landings and was not qualified to fly any type of aircraft in the group, I got the COs and Farrington in a conference and we decided that it would be a good idea to have a squadron engineering officer, or his assistant, present at all times with me at the Air Officer's Station in Fly One if that squadron had planes flying. As an Air Group we had very little time to "shakedown" and prepare for wartime operations, and we failed our Operational Readiness inspection in Hawaii on the way to Korea. Realizing our sad state of preparedness, Captain Dufek prevailed

Late in the war, the Navy adopted the F4U Corsair for carrier operations. (United Technologies Archives)

Left: Fleet Admiral Chester N. Nimitz addressing guests and crew at recommissioning of **Antietam** on hangar deck January 17, 1951 at the U.S. Naval Air Station Alameda, CA. Seated L to R are: Capt. George Dufek, C.O. **USS Antietam**, VAdm. G. D. Murray, ComWes Sea Frontier, Mrs. Nimitz, and Cap. J. P. Heath, Com Alameda Group PacResFlt.

Below: The new look — the rebuilt **Antietam** with her angled deck. (Naval Aviation Museum)

upon Admiral Radford (Commander-in-Chief of the Pacific Fleet) to come aboard and give the crew a pep talk. The Admiral made his speech brief and to the point, and it went like this: "I know you boys are not ready for combat in Korea, but we are going to send you out anyway. The fellows you are going to relieve have been fighting for over six months, and they are ready to come home, just as you will be when you have been fighting that long. After you have been on the firing line for a week, you will be as good as the best of them. Bon Voyage, and Good Luck!"

What the Admiral said was true. We had not been on the firing line a week until we could catapult 30 jets in the air in 5 minutes, and could have more than 60 planes in the air in less than 10 minutes. We had a launch or a recovery every 1½ hours, and the air department would be at flight quarters as long as 22 hours for 3 days in a row. Then we would spend 1 day replenishing our bombs, rockets, ammunition, fuel, and supplies. We mantained this punishing schedule for 8 weeks before we went to the Yokosuka Naval Base in Japan for repairs. We were there for 10 days as the entire teakwood flight deck had to be replaced, and the Japanese did it all with hand tools. We also had them erect an 11-foot nylon barrier about half way down the flight deck to stop any plane from going into the stack of planes forward. The intensity of operations required us to arm and refuel planes at the same time we were recovering them. We had one serious accident during the first 8-week period when a jet made a hard landing that wiped out his nose wheel. When this happened, the jet could not activate a Davis barrier to arrest it, and it plowed into the planes forward that were being refueled and rearmed for the next launch. The impact of the jet

*"Operation Pinwheel" was used to guide **Antietam** into port. Aircraft were placed on four corners of ship, prop blast effectively turned ship without use of tug boats.*

*Commander Robert F. Farrington, USN, Commander of Carrier Air Group Fifteen, assumed command of the group when it was commissioned in April 1951 and continued in command during its eight months tour of duty in the Korean Area on board the **USS Antietam**.*

New Slant on Carrier

CANTED-DECK CARRIER IN SEA TRIAL—The aircraft carrier Antietam, first to be equipped with a canted flight deck, undergoes evaluation trials off the coast of Virginia. The destroyer Corry is alongside taking on fuel. The angled runway (upper right) is designed to make landing safer and easier for jet planes. This picture was released by the Navy at Washington yesterday. (U. S. Navy photo via A. P. Wirephoto.)

This wirephoto released Oct. 1, 1953, 2 months before night photo from blimp.

*Commander James H. Newell, USN, became Executive Officer of the **Antietam**, September 26, 1951, just before she set sail for the Korean Theatre. He had been closely connected with carrier aviation since he received his wings in January 1940. Prior to reporting aboard the **Antietam** Commander Newell served as Aviation Plans Officer on the staff of Commander Seventh Fleet and Staff Commander First Fleet.*

Pages From *Antietam's* Scrapbook

ENTERING DRYDOCK is tricky business, requires precision. Several tug boats and scores of men were required to get ship into position.

ANTIETAM LEAVES THE WATER

Ship goes in for repairs

Enroute to Yokosuka at one time during cruise, the Antietam sustained slight damage which necessitated underwater repairs. The ship was sent to drydock for a few days before returning to the Sea of Japan for further operations.

While in drydock, normal "In Port Routine" was carried on with one minor exception. Some difficulty was encountered in connecting the ship with sewage disposal system on the dock. As a result, during first day, Antietam's crew had to use two comparatively small lavatories off the ship. Japanese workers entered into the humor of the incident with a sign warning American sailors that "Use of Japanese head requires some skill."

HIGH AND DRY. Antietam takes on different look. Diver checks position of ship before water is pumped out, wood and concrete platforms hold ship firmly in place.

JET ENGAGED arresting wire, but too late to prevent crash into barrier.

IMPACT causes plane to nose over. Crash crew tips plane, puts crash wagon under nose.

HASTY REPAIRS are made by arresting gear crew after plane had crashed barrier.

WAVED OFF by LSO, jet whose approach was unsatisfactory applies power, skims off port side without touching down.

ARRESTING GEAR MEN ALERT TO DUCK AS JET MAKES PERFECT APPROACH.

SHIP'S COMPANY
USS ANTIETAM (CV-36)

OPERATIONS OFFICER

CDR C. E. ROEMER
Cadet Class 105-C

ENGINEERING OFFICER

CDR W. P. REULAND

AIR OFFICER

CDR F. J. HILL
Cadet Class 130-C

GUNNERY OFFICER

LCDR J. C. VANDIVER

NAVIGATION OFFICER

LCDR J. E. HADLEY

SUPPLY OFFICER

LCDR E. C. SOCKERSON

MEDICAL OFFICER

CDR V. A. BEUERMAN

DENTAL OFFICER

CDR M. J. BRANDT

Rear Adm. George Dufek in his better-known later role as an Antarctic leader, third from right. (USAF)

A runaway F9F on the carrier ***Princeton*** *off Korea, 18 October 1950, has just missed the arrester wires and is about to crash into the barrier, which failed to stop it. However, the pilot survived the smash up on the unangled teakwood deck. (USN — LCDR Scott Jones)*

into the parked planes decapitated the pilot, jammed the throttle in the full-forward position, and knocked three planes over the side. The impact also killed three other people and injured nine others, while damaging several aircraft that remained on deck. We were fortunate that we did not have a fire, or we could have lost more than 100 lives and a squadron of planes.

During our stay in Yokosuka I was involved in an amusing incident with my catapult officer, Nimrod Tucker, who was a World War II reserve pilot called to active duty for the Korean War. Nimrod and I had closed an Air Force Officer's Club which was miles from our ship and the Navy Yard, so we began walking at about 1 a.m. in the hope that we would find a Geisha house. We walked about an hour and came to a fire station, and it occurred to us that this was just the type of transportation we needed to a Geisha house. We could not speak a word of Japanese, but after about ½ hour we conveyed the idea to them that we were many miles from our ship and we wanted them to take us to a first-rate Geisha house in one of their fire engines. Once they got our message, they thought the idea was hilarious, but they had to go to the Mayor's office to get permission to test one of the engines. Whoever the authority was thought the request was a bit unusual at such an hour in the morning, but granted permission, and we were off and running. The Geisha house was a nice one, and we enjoyed the hot saki, a ''hotsie'' bath, and some well deserved sleep before the girls awakened us in time to get a taxi back to the ship for the 8 o'clock muster. I do not know where Nimrod is today, but he was a fine catapult officer for the *Antietam,* even though the pilots affectionately dubbed him "Coldshot" Tucker.

When we finished our Korean tour, the commanding officers

When we finished our Korean tour, the commanding officers of the four squadrons took them back to New York, Chicago, and Denver, and they considered their year of active duty so rewarding that they tried to join the regular Navy. However, the Navy turned down their applications, and they had to go back to civilian life and start all over. They naturally did not want to be affilliated with the Reserves any longer and they all resigned their Reserve commissions. The only one of the four commanding officers that I have kept in touch with is Roland Kenton, who went back to Larchmont, NY, where he started a highly successful business; and he lives there with his wife, Doris.

After completing my tour as Air Officer of the *Antietam,* I was ordered to duty as Fleet Information Officer for Adm. Lynde D. McCormick and the Atlantic Fleet. While in this billet I began to stress the need for a canted, or angled, flight deck for all carriers, and my old ship *Antietam* became the first carrier to have one installed on a temporary basis. The ship began testing all types of aircraft and the effectiveness of the angled deck under all types of conditions. We even scheduled a night photographic mission of the carrier in operation off the coast of Norfolk, VA. A series of photo-flashes were installed aboard the carrier, and they were synchronized to be set off when a jet was being catapulted while another jet was landing on the angled deck. This instant shot was to be photographed by top cameramen from the invited media who were aboard a Navy blimp which was hovering above the carrier. On the appointed night, everything was scheduled to happen at 2000, but the blimp had to be in the exact position at the time the plane was catapulted, so the catapult officer actually got his signal from the blimp. Everything went off as scheduled, and the photo was such a success that it made the cover of *Life* and *Newsweek*. The inside stories stressed the importance of the angled deck to the safety of carrier aviation, and every carrier that was built from that time on had an angled deck.

The Grumman F9F-1 Panther carrier jet. (Grumman)

After my tour on the staff of Admiral McCormick as Fleet Information Officer, my future tours seemed rather tame and left much to be desired, so I began to plan for the day when I would complete 20 years of service. I had a BS degree from Indiana University School of Business Administration, with a major in Marketing, and I had never had an opportunity to use this education, so I began to plan another career as a stockbroker. During my last three years of Naval service I enrolled in two correspondence courses and completed them before I retired. While scouting various brokerage firms around the country, I learned that Merrill, Lynch was about the only firm that had a formal training program, so I decided to aim for employment with that firm. I made an application for employment in the Jacksonville office and took an aptitude screening test there in the early part of 1959. Once I passed the screening test, I was interviewed by the managers of the Atlanta and Jacksonville offices and was told that I could go to work in either one.

After my tour with CincLantFleet, I was Executive Officer of the NROTC unit at Georgia Tech in Atlanta. We loved the city and knew many people there, and I believed the prospects for a novice broker were better in that city, so I went to work for Merrill, Lynch there in September 1959, almost two months before I was retired from the Navy. I was using up accumulated leave while I was studying to be a broker. I worked there for more than 15 years, specializing in Government Securities and Institutional type business. I became one of the firm's most productive salesmen in the southeast and rose to become Assistant Vice President and an officer with the firm before my retirement in January 1975. That December I moved to Floral City, Florida (we have facetiously called it "Wrinkle City"), and have lived here ever since. Many of my friends in Merrill, Lynch have asked me what I do with all of my spare time in retirement. For the last three years I have been doing research and gathering stories about some of these "Early Birds." It was difficult to fit the stories into any logical sequence, and thus I started using the random approach.

One of these birds has come a long way since he became involved in flying for the Navy. Harvey "Hippity" Hop owns a Lear Jet and runs a jet charter business out of Ft. Lauderdale, FL, which he calls Hop-A-Jet, Inc. He has averaged 125 pilot hours a month over the last seven years, and he claims to have more Lear Jet pilot time than anyone in the world. Harvey has amassed a total of more than 35,000 accident-free hours to date and is still flying strong. If you need a charter, call Hop-A-Jet at 305-563-6633, and he will pick you up, day or night, to take you anywhere his Lear will fly. Harvey was 65 in July 1982 and is the only one of these "Early Birds" who is still actively flying for a living. He and his wife, Miriam, live in Ft. Lauderdale.

Classmate Jack Coley is still flying, but not for a living. On their Christmas card in 1982, he and Lois wrote that they were going to buy a new airplane and might fly to Florida to see us. While Jack was flying for the Navy, he became an expert in polar aviation. He flew for Admiral Byrd on one of his Antarctic expeditions, and was with his successor, Adm. George Dufek on several of his Antarctic trips. Jack is the only cadet-trained pilot who has flown over both poles. We were not only classmates at Pensacola, but were pilots in the same squadron at Pearl Harbor, Kaneohe, and the South Pacific. We were even sent back to Jacksonville at the same time in January 1943, but Jack went into training as a PV pilot while I remained a flight instructor in PBYs there. After Jack retired from the service he worked for Lockheed Aircraft in Los Angeles for more than 20 years before he and his wife, Lois, retired from that firm and moved to Chicago Park, CA.

The only other classmate I know who is still alive and flying his own plane is John F. Jonas. After more than 20 years of flying for the Navy, John went to medical school, served his internship, and became an M.D. He has a very successful medical practice in Forrest, VA, today and flies his own plane to medical conventions and cadet reunions.

My good friend Jack L. Poteat, Class 150 C, flew for the Navy during the war years and then went to work as a check pilot for the FAA, in Atlanta, GA. He retained his active reserve status as a weekend warrior and served two weeks of active duty each year until he was retired as a Captain in the U.S. Naval Reserve. Jack checked the proficiency of airline pilots in all types of jet and propeller driven planes for more than 30 years out of Atlanta and Dallas before retiring from the FAA in 1974 with about 40,000 hours of pilot time. He has not lost his interest in aviation as he and a friend are building a small fighter plane that they plan to use personally for business and pleasure. Jack is still playing golf from the wrong side, when he is not dabbling in real estate. He and his wife, Flora, are enjoying retirement in Atlanta, and in their condominium, "Spanish Trace," at St. Augustine Beach, FL.

Walter A. Hibbs, Class 81 C, was one of the first cadets to get his wings and become an aviator, and he wrote a lot of letters to his parents and his girl friend, Francis, while he was at Pensacola. His parents and Francis thought so much of his letters that they preserved every one of them for him, and he still treasures them today. Walter had the idea that he would some day write a book about his experiences, but he had not done so by the time he received my letter which invited stories from pre-World War II cadets, and he decided to lend them to me for background material. Walter also loaned me his copy of the 1935-36 yearbook (Flight Jacket) and his copies of the flight and ground school syllabuses. I am deeply indebted to Walter for loaning me this valuable material, and I hope that every person who reads this book will appreciate his kindness and generosity in giving me a helping hand. After Walter served out his four-year contract with the Navy, he became a pilot for PanAm and flew as a Captain for more than 30 years until he retired in April 1972. He and his wife Francis are living in Carmel, CA.

Frank J. Hill

Age - 41 Height 5' 7 Weight - 155

Marital Status - Married - 2 Sons - Ages 14 and 9

Present Assignment - Public Relations Director and Service Information Officer Naval Air Station, Jacksonville, Florida

Availability - Retiring as Captain U S Navy on 1 July, 1959 and available after that date. Prefer to arrange interview schedules prior to 1 April, 1959

Home Address - 1390 Belvedere Avenue
Jacksonville, Florida

Home Phone - EVergreen 9-6327

The first page of the author's resume as he prepared to leave the service and transfer to Merrill, Lynch.

Many of the first-year cadets became airline pilots once they finished their Navy contracts and collected their $1,500 bonus. Eight other pilots besides Hibbs went with PanAm in 1939, and at least six of these are still living today as retired Captains from that line, including my first flight instructor, Bart Randolph, who lives with his wife, Maggi, in Harrison, AZ. At least one of the three pilots in 81 C who went with TWA is still alive today. He is a retired Capt., George S. Felt, who lives in Hyde Park, NY. Two pilots in 81 C elected to make the Navy a career. I had the pleasure of serving with one of them, Vincent F. Casey, as his Executive Officer on Guam, and I must say that he was one of the finest cadet-trained pilots I ever met. Vince is retired and living with his wife, Marjorie, in Cape Coral, FL. The other cadet in 81 C who made the Navy his career is John J. Lynch; and he became the first cadet-trained pilot to reach flag rank. He is a retired Rear Admiral living in Washington, D.C., with his wife, Virginia. Wherever the pilots in 81 C went, they gave an outstanding account of themselves, and they set the highest standards for the cadets who followed them.

My good friend and squadron mate, Charles Fountain Willis, the same fellow I saved from drowning in the tub in our Kaneohe house, came out of World War II a well decorated pilot and served a year as a Navy test pilot. He then became an entrepreneur in a number of businesses involved with aviation, including an Air Freight line, the Willis-Rose, Corp., the Teterboro School of Aviation, and Aircraft Service International, Inc. During the 1952 presidential election campaign, he co-founded "Citizens for Eisenhower," which assured Ike's election as president. He was rewarded with the appointment of Special Assistant to the President for Patronage and Aviation Matters, a short time after the inauguration. After about two years in this pressure cooker job, Charlie wanted to get back into flying and get the hell out of Washington. When he bought control of Alaska Airlines and became its President and Chief Executive Officer, this satisfied both requirements. He held this position for more than 15 years, and under his leadership, Alaska Airlines helped Alaska become the 49th state. Also while he was President, he opened the door to Siberia, as his airline made the first charter flights there. He is no longer actively flying, but you give him 15 minutes in the cockpit of almost any type of plane and he will take off. He has done some aviation consultant work in the Miami area in the last few years, but spends a lot of time on the waterways around Ft. Lauderdale in his 25-foot sloop with his wife, Marion. He is now a founder and President of Pacific Aerolift, Inc., a cargo airline based in Hawaii.

Another good friend and squadronmate, and one of the original five in our Kaneohe house, is George Poulos, Class 132 C. After the war, George decided to make the Navy his career and accepted a regular commission as a Lt. Commander, but he became disenchanted with the lack of flying activity it offered during peacetime, and he decided to resign in 1947. He returned to his father's vineyards for about a year, but found that he could not stay away from aviation, so he went to work for Lockheed in Los Angeles, and has been with that firm ever since. He now lives with his wife, Peggy, in La Cãnada, CA.

Carlton H. "Dopey" Clark, Class 122 C, came back to the U.S. after his 3-year ordeal in Japanese prison camps and gave the Navy 22 more of his best years, ending as the Commanding Officer of the Naval Air Station, Agana, Guam. After he retired from the Navy, he found a job doing exactly what he wanted to do. He is the Public Relations representative for the Southern Sash Company in Montgomery, AL. In this position he has the responsibility of flying and maintaining the company aircraft and keeping the company yacht standing by in Pensacola, ready to take customers fishing. He also has charge of a company condominium where customers can live it up if they wish to remain ashore, or they do not want to fly. Carl's nickname at Pensacola and in the squadron was "Dopey," but you only had to be around Carl a few minutes to realize he is anything but that. He might not have appeared to be paying attention, but he always had the word and was sharp as a tack; now he seems to be having fun doing what he wants to do while he is getting paid for it. Carl lives in Pensacola, and rarely takes a vacation, because he doesn't need one.

Squadronmate Charles P. "Muck" Muckenthaler gave the Navy 30 of his best years as a pilot before he retired in San Diego and went into business for himself. As a Financial Planning Consultant, he keeps busy advising his old Navy buddies what to do with their money.

Squadronmate Maurice "Sully" Kauffman flew for more than 20 years for the Navy before he retired as a Captain in the U.S. Naval Reserve. Then it did not take him long to find out what he wanted to do. Sully had always been interested in yachts, so he became a Los Angeles Yacht Broker. If you need a yacht, he can find you one, but if you have to worry about the price, you don't need one. Sully was in Class 106 C, and he and his wife, Trudy, have attended all three cadet reunions.

Squadronmate and classmate Bob Corlett started flying on the same day that I did in St. Louis, MO, in June 1939, and we had identical assignments in the Navy for the first five years. Bob and his wife, Eleanor, were smart enough to buy a house in Coronado during their first tour of duty at North Island. They have raised a wonderful family of three boys and two girls.

My good friend and classmate Frank L. "DeLo" DeLorenzo and his wife, Helen, showed a high degree of intelligence when they built a lovely house near the 18th hole of the Pensacola Golf and Country Club during their first tour there. Delo and Helen gave the Navy 30 of their best years before retiring. They have been active in making preparations and planning for the three pre-World War II aviation cadet reunions we have had to date (two in Pensacola and one in San

Diego), and they are getting ready for another one in Pensacola in November 1983. They have compiled a list of about 1,200 of these ex-cadets who are living today, and have kept an up-to-date record of their addresses. DeLo and Helen have received invaluable assistance from Dick Rossi, Class 134, and his wife, Lydia, and Bill Graham, Class 136, and his wife, Dorothy, in hosting these reunions. Rossi was a Flying Tiger when World War II began, and was a pilot for the Chinese National Aviation Corporation flying vital supplies into China over the Hump, from the time the Tigers disbanded until the war was over. He and Lydia live in Fallbrook, CA. Bill Graham is an author who wrote *Yellow Birds and Silver Eagles*, a book about his experiences as a flight instructor in the Stearman primary bi-plane trainer, "Yellow Bird" or "Yellow Peril." He says it took him five years and 4,000-man hours to write the book, and he did it all in between working hours as President of a lumber sales company. He and Dorothy live in Northbrook, IL. If it had not been for these people and their aviation cadet reunions, I could not have written this book.

There was another in the first flight class who wrote about his experiences while going through flight training. He wrote the story titled *Wings of Gold* published in *Redbook* in the late thirties. Boone Guyton was also involved in flight test work after he finished his four-year contract with the Navy, and he helped to develop the F4U gullwing Corsair. He is now retired and living with his wife Susie in Woodbridge, CT.

My good friend Gordon Forbes, Class 135 C, spent 20 years flying for the Navy before he retired and did some writing. He had a book published titled *Goodbye to Some*. It covered his wartime experiences in PBYs, B-24s, and Privateers. Gordon has played a trumpet for more than the 44 years I have known him, and he wrote to me recently that he has had his own combo in Aspen, CO, playing five and six nights a week. He was always available in the barracks at Pensacola for a "jam" session with Ed Stepanek on accordian, Bob Corlett on trombone, and yours truly on the sax. Among Gordon's other accomplishments, he was a Captain for American Airlines for many years before he retired and went back to blowing his trumpet.

Another good friend and fraternity brother, Donald L. "Pop" Korn, Class 128 C, was a great influence on me to become a naval aviator. Don was several years ahead of me at Indiana University and had graduated from Law School before he went to Pensacola. As a student, he always had several things going for him, and he was a super salesman. He was involved in several enterprises at Indiana while I was a student, and he made more money than most workers who had a full-time job. At the beginning of the war Don was caught aboard the cruiser *Raleigh* at Pearl as the off-coming Duty Officer, and later on he was involved in the Navy's secret Drone program with Mark Wood and "Rosy" Lyons. Mark went to the South Pacific as Commanding Officer of his unit and made about 50 sorties with drones against Japanese installations. Don did not make any overseas deployments with his special task unit, as Rosy Lyons and his unit deployed in England met with disaster. When Joe Kennedy, Jr., and Bud Willey were destroyed before they could bail out of their torpex-laden B-24, the drone program came to a halt at the insistence of Joe Kennedy, Sr. After the war was over, Don organized a small air-cargo operation called GIMACO (G. I. Mutual Aviation Company), but it was vastly undercapitalized and went "belly-up" after a few months. From this sad experience Don became a laborer in Tulsa, OK, as a "Tanky" with a large gas storage company at 80¢ an hour. After a year of this, he was offered opportunities for advancement, and in 1955 he consumated a deal to buy a division of the company and moved it to Supulpa, OK, where it became Tresco, Inc.; and he is still running that company there today.

Gregory "Pappy" Boyington, Class 88 C, was a Marine flight instructor at Pensacola in 1941 when he resigned his commission to join Claire Chennault and the AVG group in China. He was an Ace with six planes to his credit when the AVGs disbanded in July 1942, and later on in the Pacific he shot down five planes on his first mission in his Black Sheep squadron. He had a total of 28 planes to his credit when he was shot down, and he was awarded the Medal of Honor. Years later Greg became an author when he wrote *Baa Baa Black Sheep*, and he became a technical advisor for the famous television series with the same name.

James H. Howard, Class 109 C, was another AVG and another Medal of Honor winner. He was an ace with the AVGs, collecting $500 for each confirmed plane shot down, and was offered a commission in the Army Air Force as a Major with a squadron of his own. However, he was suffering from Dengue Fever and decided to turn down the offer and go back to the States to regain his health. After several months of recuperation, he accepted an Army commission as a Captain in the 354th Fighter Group, which was sent to England in the fall of 1943 to fly P-51 Mustangs. One day in January 1944 he led a group of 50 fighters to cover B-17 and B-24 bomber groups at 25,000 feet. On this mission he shot down three different types of German planes (a twin-engine ME-110, an ME-109, and a Focke-Wulf 190). He expended all of his ammunition, but continued fending German fighters away from the bombers until he had to return to base because of lack of fuel. He received the Medal of Honor for this day's operations and was the only fighter pilot in the European Theater to receive such an award. Later he was an ace against the Japanese, as well as the Germans. Jim eventually took over Command of the 354th fighter group when his CO, Ken Martin, was shot down and became a German POW. When the war ended, his group had the greatest number of aerial victories (701) and one of the lowest pilot loss ratios. He left the service after the war, and eventually established his own company, Howard Research Corp., in Arlington, VA, which was merged with Con-

The last official portrait, 1953.

trol Data Corp. in 1965. Jim remained active with the Air Force Reserve and is a retired Brigadier General, living in Florida.

Henry C. Bridgers, Class 82 C, was a shipmate with me on the staff of Adm. Lynde D. McCormick, Commander-in-Chief, U.S. Atlantic Fleet in 1952-53. He flew for the Navy for 30 years before retiring to his hometown of Taroro, NC, and the business of caring for the family farm. Henry also actively pursued his interests in writing and historic preservation. He spent much time in the South Pacific during World War II, and in 1977 he decided to revisit the area. He was one of six people who arranged a tour through Veterans International Tours of Brunswick, ME, and they spent more than a month visiting 20 South Pacific islands and flew more than 18,000 miles. Henry has had several articles published, but his account of the South Pacific, published in the spring issue (May 1980) of the Naval Aviation Museum Foundation magazine, was his finest effort. Henry informed me that he had several other articles in the works, but unfortunately, he had a heart attack and passed away shortly after the Foundation article was published.

Alden Cobb Barnard, Class 81 C, authored an article about an incident that happened at Pensacola while he was a cadet. The date was 14 January 1936 and the event was the landing in Pensacola Bay of the world's largest flying-boat, the French *Lt. de Vaisseau*. Two of these boats had been built by the French at a cost of more than $1,000,000 each, and that was a ton of money in 1936. The plane had completed a 2,000-mile, 20-hour flight from Martinique to reach Pensacola, and it was quite a sight to see. It weighed 37 tons, could carry 70 passengers and crew, was 104 feet long, had a wingspread of 162 feet, and had 6 engines putting out 850 horsepower each. The plane was moored to a buoy in the bay about 100 yards from the ramps of training Squadron Four. During its first night at the buoy, a severe squall passed through the area and flipped the plane on its back, and it sank in about 25 feet of water. LCdr. J. R. Bonnot and his crew of six were a tearful group when they learned that their million dollar aircraft would be loaded aboard a freighter and shipped back to France, where it would have to be completely rebuilt and reassembled.

My good friend Roger Crowley, Class 106 C, became a Republican politican in New Hampshire after he retired from the Navy. He campaigned for Governor twice and came within a few thousand votes of being elected in 1978. If Rog could have run one more time, I believe he would have been Governor. Unfortunately, he came up short in the race with time, and he passed away from a heart attack before he could run again.

One of the most famous cadet-trained Marines, Joe Foss, Class 150 C, was retired as a Brigadier General, with a total of 28 planes shot down, before he decided to become a politican and Governor of his state. Like the outstanding pilot he was, he became one of the finest Governors the state of South Dakota has ever had. He is now fully retired from active politics and lives in Scottsdale, AZ.

Of the more than 1,200 cadet-trained pilots who are alive, who were trained before World War II, and who want to be affiliated with our rather loose organization, more than 150 are retired airline Captains. Some of these pilots were AVGs and elected to fly for the Chinese National Aviation Corporation after the Tigers disbanded in July 1942. After the war they pooled their knowledge and resources to become entrepreneurs and organize the Flying Tiger (air freight) Airline. Their knowledge of how to get freight where it is supposed to go in a short time has paid off handsomely in this venture and has made wealthy men out of all of the original stockholders. There is only one thing I would change about their advertising slogan *(If it's not on time, it's on us)*. I would change this to, *If it's not on time, it won't cost you a dime.*

The majority of the first-year cadets (Classes 81-88 C), who finished their four-year contracts with the Navy and wanted to stay in aviation, were hired by the airlines, and PanAm got the "lion's" share of these fine pilots. PanAm needed these Navy cadet-trained pilots because they were the only ones who were learned in overseas navigation at the time, and all of PanAm's flights were overseas. TWA probably hired the second largest group of cadet-trained pilots; after 1945, many of their flights were overseas. However, some of the other early airlines were not to be denied the privilege of hiring first-year cadet-trained pilots (from 83 C), as Eastern hired Charles O. Rogers, Delta hired William Bails, and United hired Gerald L. Pentz, the first cadet to join the Caterpillar Club. They have all retired from their respective airlines with almost 100 (total) years of flying and about 75,000 hours of pilot time.

Another one of the first-year cadets that responded to my letter about writing a book, was Page W. Smith, Class 86 C, who searched for Ameila Earhart in July 1937. He was a PanAm captain for more than 30 years before he retired to become a gentleman farmer in the eastern shore of Maryland in the summers and spend the winters in West Palm Beach, FL.

Two of the young Ensigns who accompanied me on the *Matsonia* in August 1940 from Los Angeles to Hawaii and our first duty station at Pearl Harbor were Neal Lytle and Louis Lucas, both of Class 131 C. Neal, Paul Hart, and I drove across country together before we got to Los Angeles, but we met Lucas, and "Cloudy" Riley there to board the *Matsonia* for the voyage. Paul Hart was killed in the early days of the war, and Cloudy and I elected to make the Navy our careers. Neal Lytle found that love was stronger than regulations when he found Monique, and they were married long before his agreed two years of bachelorhood expired. When this fact became obvious to his Commanding Officer, Neal resigned his commission and began flying for TWA. After more than 30 years as a Captain with that airline, Neal is retired and living with Monique in Stuart, FL. When Lucas finished flying for the Navy, he began flying with Aloha Airline in Hawaii. He is now retired and living with his wife, Thelma, in Honolulu. Another friend, Ercel Hart, 109 C, was a Captain with Aloha and is now retired and living with his wife, Matilda, in Kaneohe.

Twenty-four of the original AVGs (Flying Tigers) who were with Chennault in China between July 1941 and July 1942 were cadet-trained Navy and Marine pilots who are on our reunion roster today. When one considers the number of pilot hours in this group, and their current age, their survival rate is incredible. Almost all of the pilots were Aces when the group disbanded in 1942. During the period from 1 October 1941 to 1 April 1942, the group shot down more than 300 Japanese planes with the loss of only 6 Flying Tigers, and they destroyed almost as many planes on the ground. They covered cadet flight training classes from 88 C, Greg "Pappy" Boyington, to Class 157 C, C. H. "Link" Laughlin, who is a retired Colonel in the Marine Corp Reserve, living in Coral Gables, FL. Both Colonels are authors of some renown. This group includes four retired officers who defected from the Navy to the Army Air Force during the war — Jim Howard, 109 C (covered in an earlier section of this chapter); David Lee "Tex" Hill, 121 C, a retired Air Force Colonel and entrepreneur living in San Antonio; TX; Edward F. Rector, 129 C, a retired Air Force Colonel living in Alexandria, VA; and Charles H. Older, 129 C, a retired Air Force Lt. Colonel. Although these gentlemen failed to shoot down any planes for the Navy, they gave the Japanese hell in China and their opponents hell while they were in the Air Force. The combined number of planes shot down by these four is 66 plus. Older is presently a Superior Court Judge in the Los Angeles area, and he tried the famous Charles Manson case. Robert "Buster" Keeton, 121 C, flew for PanAm for more than 30 years and claims to have more (25) around-the-world flights than any other cadet-trained pilot. Two other AVGs, William H. Fish (120 C), and Robert Layher (132 C), are also retired PanAm Captains, and these three pilots have a total of more than 75,000 hours in the cockpit. Most of the group that stayed in China flying the Hump came back to the U.S. after the war to establish the Flying Tiger Airline. One of the few cadet-trained pilots that remained loyal to the Navy is Noel R. Bacon, Class 129 C, who is a retired Captain, operating as a private mortgage lender in Boca Raton, FL. Of the remaining AVGs on the roster (24), they were all involved in active flying or in businesses closely related to aviation until they retired, and they were all highly successful.

William "Tiger" Tanner, Class 129 C, the same pilot who sank a mini submarine about 0630 in the morning of 7 December 1941, gave the Navy more than 25 of his best flying years before he retired as a Captain. He was not happy out of aviation, so he went to work for Northrop and is holding down an important position with that firm today. He and his wife, Nita, are living in Los Alamitos, CA. Tanner's good friend and classmate, Mark Wood, drove across country with their good friend Grover Gilmore in August 1939 to begin flight training. They all three live in California, have remained the best of friends through the years, and have had several private reunions. Wood was with a super-secret radio-control group, masquerading as a utility squadron, on the island of Maui when Tanner dropped his depth charges on the Japanese mini-sub, and two hours later he was ordered to take his planes out 200 miles from Oahu in search of the Japanese fleet. Luckily they did not find anything. Mark was later Commanding Officer of a Special Air Task Force that deployed to the South Pacific using radio-controlled drones. His group made about 50 sorties with drones before the program was halted. After the war, Mark went to law school, became a lawyer, and is now a Municipal Court Judge in the South Bay Judicial District in Torrance, CA. He and his wife, Ebba, live in Manhattan Beach, CA.

One of my good friends, Hugh "Buck" Rogers, is a retired airline Captain living within ten miles of me here in Citrus County, FL. Buck was at Pensacola when I arrived in 1939 and was in Class 120 C. After he got his wings, he was assigned to a PBY squadron in Coco Solo, Panama. Our paths did not cross again until we were shipmates undergoing flight training in B-24s and Privateers in Hutchinson, Kansas, in 1944-45. We finished our Privateer training at Camp Kearny (now Miramar) just north of La Jolla, CA, had our "transpac" flights to Kaneohe at about the same time, and went on to the Pacific to command Privateer squadrons. We had "going away" parties at every stop on the way out in August 1945. After the war Buck flew many years as a private pilot for Standard Oil of California before he became a Captain for Airlift International. He is now retired and living with his wife, Yoshiko, in Inverness, FL, and we have had many "coming back" parties since I have come to Florida. The most recent one was his 67th birthday party on 19 January 1983, and before that at Thanksgiving we had a get-together with his good friend and squadron mate, Tom Sutherland, Class 119 C. Tom and his wife, Helen, are living in Morristown, Tennessee.

George Gay began his flight training at Opa Locka, FL, on 15 February 1941 as a Seaman 2nd Class. I do not know what flight class he was in, but his first assignment after he got his wings was Torpedo Squadron 8, which was at Norfolk, VA. After the war started, his squadron was assigned to the USS *Hornet* along with Fighting, Scouting, and Bombing squadrons, to become Air Group 8. George was aboard the *Hornet* when it delivered Jimmy Doolittle's B-25s to their launching point for bombing raids on the main Japanese island of Honshu, and he was involved in the Coral Sea battle, but he did not become famous until he learned that he was the sole survivor of his squadron at the Battle of Midway. After the war he began flying for TWA and was a Captain for more than 25 years. He had

25,000 hours with the airline before he retired to Naples, FL. After that, George began to get his stories together and decided to write a book, *Sole Survivor*. The book was copyrighted in 1980, but I did not receive my autographed copy until January 1982. It is somewhat unique in that he financed the publication of the book himself by a house he called Midway Publishers. George was called to Hollywood as a technical advisor during the filming of the movie *Midway*, where his part in the battle and his survival was depicted. When I asked George for advice about financing the publication of this book, he advised against it, but said if I wished to go this route he would help me. He also advised me to make it a paperback rather than a hard cover if I financed it myself, because publication of hard covers was too expensive. When I received this advice I turned to Dr. Robin Higham, at thc *Aerospace Historian* editorial offices at Kansas State University.

Classmate Houlder Hudgins flew for the Navy more than 20 years before he decided to hang up his flying suit and do some ground work with the engine overhaul and repair department at the Jacksonville Naval Air Station, where he is still working today. He and his wife, Margaret, live in Orange Park, FL, but I think "Hudge" is about ready to retire from the business and become a gentleman farmer on his Virginia farm. Squadronmate M. R. "Kaiser" Uebelacker relieved George Poulos as the squadron "Yardbird" in December 1940 at Pearl Harbor and went on to give the Navy some of his best flying years before retiring to Pensacola and working as a civilian in the Naval Air Station engine overhaul and repair department there. "Uebe" married Marguerite, a Pensacola girl, as soon as he got his wings, more than a year before the agreed two years of bachelorhood expired, but the Navy was never officially notified, so he remained in the service until his retirement. Like Hudge at Jacksonville, he has been in the O & R business for more than 25 years, and I expect he is ready to retire from that position, though I imagine he will stay in Pensacola.

Classmate John F. "Jack" Davis was in a PBY squadron at Pearl Harbor when the war started and went on to the South Pacific during the early days of the war. His squadron finally retreated to Perth, Australia, where he met and married his wife, Mary Frances. He retired from the Navy in 1961 and they moved back to Australia, where Jack is an executive with a firm that is exporting Solar Hot Water Systems to Hawaii, San Diego, and Los Angeles. Two other good friends who were PBY pilots and went to Australia fell in love and married Australian girls. Dennis M. Szabo, Class 107 C, married his Rita in 1944, and Carl "Pete" Schmuck married Joan in 1942. I have lost track of the Schmucks who may be living in Australia, but Szabo lives in Atlanta with his daughter Denise and has done so since Rita passed away in 1963.

During our last reunion in San Diego in 1981, I saw my good friend Arthur L. Jacobson, Class 121 C. I had not seen him since we had been shipmates in training at Hutchinson, Kansas, and Camp Kearny, preparatory to taking command of Privateer squadrons in the Pacific. Art and his wife Lois own their own airplane and took turns flying it down to San Diego from Mercer Island, WA.

It was not until after the second reunion in Pensacola in November 1979 that I decided to write a book about this group of aviation cadets (Navy and Marine) who were trained as pilots before World War II began. I drafted a letter, had it printed, and mailed it to more than 1,000 for whom I had addresses. In the letter I gave a general idea of the areas I wanted to cover and welcomed stories about their lives that they thought were significant. The response from more than 100 was very gratifying, and this encouraged me to continue the project. Many of my good friends down through the years received their final check long before my letter went out and their stories will remain untold. Many others did receive the letter and did not bother to answer it, so their stories likewise will be unheard. However, I did have some first-hand information about some of these fellows, which I have tried to include anyhow.

At the 1977 reunion, Crown Center, Kansas City — (above) Hill, Clark, Willis, Rief; (below) Willis, Coley, and Ricketts.

Wherever these "Early Birds" are, wherever they have been, and whatever they have done, they have given a good account of themselves, and I expect they will continue to do so until the Lord taps them on the shoulder for that final check.

Many of my good friends received that check early in life. John Fusek, Class 135 C, was a Delta Upsilon fraternity brother at Indiana University. When he finished flight training he was retained as a "Yellow Peril" instructor and sent to the new Naval Air Station in Jacksonville to train students there. After the war he became an Eastern Airline Captain flying the New York to Miami route, but he had to retire from that position in the mid-fifties due to ill health, and he lost his bout with cancer less than six months after he retired. Another good friend, Bill Chester, also in Class 135 C, was sent to Jacksonville as a "Yellow Peril" instructor with Fusek. Bill made the Navy his career for almost 30 years before retiring as a Captain in Jacksonville. He attended the first reunion in 1977 at Pensacola, the first time I had seen him since I received my wings there in July 1940. We had a great time rehashing the past, and the great double-dates we had with young lasses there, but that was his last reunion, as he lost his bout with cancer in 1978.

Another good friend, Tom Hillis, Class 105 C, received his final check early. Tom was a PBY pilot in VP-14 at Kaneohe on 7 December 1941, and like many of us was awakened by Japanese Zeroes strafing the BOQ before 8 o'clock that morning. When Tom came back from the Pacific, he went into twin-engine Lockheed PV flight training and became an instructor at Cecil Field while I was an instructor at Jacksonville. He met and married his wife, Elaine, about the same time I met and married my wife, Grace, in 1944, and Tom and I were shipmates at Damage Control School in San Francisco in the spring of 1946 before going aboard carriers. Tom and Elaine found that they could bear no children, so they adopted triplets in 1949. They were about the same age as our young son, Tim, and we saw each other frequently when we were all in the Norfolk area from 1952 to 1954. Tom was Operations Officer of the Naval Air Station, and I was Fleet Information Officer for the Atlantic Fleet. We had a mutual friend who was Commanding Officer of a Patrol squadron in Trinadad, and he knew how much Tom and I liked good rum (especially Anejo without customs duty). Everytime our friend had a plane coming to Norfolk, he would see that there was a case of rum marked for Hillis and one for Hill. One day, shortly after a plane came in from Trinadad, Tom went from his office down to the beaching crew's headquarters and said in a loud voice, "Where's my rum?" The customs man was standing less than 10 feet from Tom, waiting to check another plane and he said, "Yeah, where is it?" Tom recovered quickly and said, "I came down to make a customs declaration and pay the duty on some rum." There were four cases involved, and Tom filled out the declarations and paid the duty on all four of them, and that was the last attempt to get any duty-free spirits into the Norfolk area. Shortly after this, my boss," Admiral McCormick, put out a Fleet memo cautioning Fleet units to make customs declarations on all goods brought into port, and Vice Admiral Stump put out an AirLant memo covering the same subject.

Tom received his final check shortly after taking off from Malta in the late fifties when his R4D crashed due to engine failure. Elaine has never found anyone like Tom and deserves a high mark for raising Cathy, Jeff, and John to adulthood by herself while working as a Librarian in Orange County, CA. Elaine attended the 1981 San Diego reunion as the widow guest of the Gilmores, and we made an overnight stop in her Tustin home on Sunday night after the reunion and got to see the children who will be 34 years old in June 1983. Another guest at the same reunion was Lil Crowley from New Hampshire, who was the widow guest of Gordon and Pat Egbert from New Orleans.

At least four of my friends, including two classmates, received the final call in 1982. They were Danny Liane, the first cadet-trained pilot to sink a German submarine after the war began (off the coast of Martinique, 10 December 1941), who lived in Pensacola, and Fred Andretta, who lived in Seattle, WA. The other two I know about were Leroy "Pete" Keith, Class 118 C, the pilot I relieved as Executive Officer at NAS Agana, Guam, in 1956. Pete gave the Navy more than 20 of his best years, shot down more than enough Japanese to become an Ace, and had a hand in sinking more than 20 Japanese ships. He was a Real Estate Broker in St. Louis, MO, where he and his wife, "Suz," lived until he received the final call late in 1982. Roy E. Reed, Class 150 C, was a shipmate aboard the carrier *Princeton* in 1946-47, and we were shipmates at the first General Line School class in Monterey, CA, in 1948 and again at the Armed Forces Staff College in Norfolk in 1950-51. We could not play golf when we were aboard the carrier, but we were on the course every day we were not in class at Line School or at the Staff College. The 1982 Christmas card from Prescott, AZ, brought the sad news from Doris that Roy lost his bout with cancer in the early part of the year.

Since the first cadet reunion at Pensacola in 1977, more than 50 of these brave pilots have received their final call, and our ranks are thinning fast. Four out of 52 have been classmates in 130 C in those years, by far the greatest attrition of any class. As far as I know at this writing, there are only 22 pilots alive today out of a total of 106 that began flight training in September 1939. Perhaps these figures are telling us something like we worked too hard, played too hard, drank too much, or we were more susceptible to disease or heart failure than cadets of other classes. Whatever the case may be, there will be more than 300 of these "birds" looking forward to another reunion in November 1983 in Pensacola, where it all began.

Final portrait — Joe Hill and his two loves: his wife of forty years and the PBY preserved in the U.S. Naval Aviation Museum at Pensacola.